THE ROYAL HORTICULTURAL SOCIETY
PRACTICAL GUIDES

PLANTS FROM
CUTTINGS

THE ROYAL HORTICULTURAL SOCIETY PRACTICAL GUIDES
PLANTS FROM CUTTINGS

Alan Toogood

A Dorling Kindersley Book

LONDON, NEW YORK, MUNICH,
MELBOURNE, DELHI

Project Editor Fiona Wild
Art Editor Colin Walton

Series Editor Helen Fewster
Series Art Editor Ursula Dawson

Managing Editor Anna Kruger
Managing Art Editor Lee Griffiths

DTP Designer Louise Waller

Production Manager Mandy Inness

First published in Great Britain in 2003
by Dorling Kindersley Limited,
80 Strand, London WC2R 0RL

A Penguin Company

Copyright © 2003 Dorling Kindersley Limited, London

All rights reserved. No part of this publication may be reproduced, stored in a
retrieval system, or transmitted in any form or by any means, electronic, mechanical,
photocopying, recording, or otherwise, without the prior written permission
of the copyright owner.

A CIP catalogue for this book is available from the British Library.
ISBN 0 7513 4890 2

Reproduced by Colourscan, Singapore
Printed and bound by Star Standard Industries, Singapore

See our complete catalogue at
www.dk.com

CONTENTS

PLANTS GROWN FROM CUTTINGS 7
An introduction to growing plants from cuttings: the history and the different techniques.

RAISING PLANTS FROM CUTTINGS 23
TOOLS AND SUNDRIES 24
ROOTING MEDIA 26
PREPARING CUTTINGS 28
TYPES OF CUTTING 30

CARING FOR CUTTINGS 49
PROPAGATING UNDER COVER 49
COLD FRAMES AND CLOCHES 52
CARE OF CUTTINGS UNDER COVER 54
THE ROOTED CUTTINGS 56
CUTTINGS IN THE OPEN 58

PLANTS TO GROW FROM CUTTINGS 61
GARDEN TREES 61
SHRUBS AND CLIMBERS 64
PERENNIALS 69
CACTI AND SUCCULENTS 76

Index 78
Acknowledgements 80

Plants Grown from Cuttings

Why Grow Plants from Cuttings?

Plants have evolved a fascinating array of strategies in order to survive and increase, and since the dawn of civilization, farmers and gardeners have used their observations to develop propagation methods in cultivation. Raising new plants from cuttings is often a straightforward process and today it is possible for everyone to make the most of the known techniques, from the child with a jamjar of water to the nurseryman with heated greenhouses.

Exploiting Nature

Propagation from cuttings began when rooted shoots or suckers were detached and replanted. This led to propagation from unrooted shoots. The great advantage of propagating plant material in this way is that the new plant is genetically identical to the parent. If you need large numbers of plants, for example tender perennials for bedding, then preparing trays of cuttings will be a much more economical method of increasing your stock than buying from a nursery or garden centre. Another plus is that if you admire a plant in a friend's or neighbour's garden, it is often possible to recreate it in your own, by taking a few cuttings and caring for them well, without damage to the parent plant.

CLEMATIS 'PERLE D'AZUR' *This deciduous climber can be propagated in various ways. Leaf-bud cuttings, for example, should be taken from softwood and semi-ripe shoots. It will take 2–3 years for the new plants to flower.*

◀ PLANTS FOR FREE *Plants like pelargoniums and fuchsias are easily grown from cuttings.*

Adapting from Nature

Propagation, and gardening generally, is easier if plants are suited to the climate and can be grown outdoors all year round. Plants that are grown outside their natural habitats generally require artificially enhanced conditions under cover, such as heat and humidity, for propagation. Success in propagation usually depends on providing a supportive environment for the plant material and later, the new plants. Understanding how plants function will help too.

Survival in the Wild

Plants are amazingly capable of survival and increase in all kinds of conditions. They have adapted to a wide range of adverse habitats, such as deserts (cacti), high altitudes where winds damage foliage and discourage pollinating insects (alpines), and water (aquatics) where the problems are different again.

All flowering plants reproduce by seeds (sexual reproduction) or by vegetative (asexual) methods. Nature has overcome the limitations of seeds by adopting asexual reproduction, producing offspring (clones) that are genetically identical to the parent.

Plants have many ways of increasing vegetatively from modified roots or stems. The simplest way they do this is by forming a closely knit mass, or crown, of shoots and buds, each of which is capable of becoming a separate plant.

Some plants can regenerate shoots or roots from growth tissue to produce new plants (runners or layers). Others form

> Cuttings are one way to exploit natural regenerative processes

special food-storing organs, including stem tubers (eg potatoes), corms (eg crocuses) and pseudobulbs (orchids). This enables a plant to survive a hostile situation, saving energy for more favourable conditions.

KEEPING WARM
An artificial environment has been created for these ivy cuttings by draping opaque plastic film over hoops on a heated bench to increase warmth and humidity around the cuttings, both factors which aid rooting.

ADAPTING FROM NATURE • 9

Gardeners have adapted natural vegetative, or clonal, reproduction to obtain plants that are always "true" to the parent. Cuttings are one artificial way to increase plants by exploiting their regenerative abilities. Clonal propagation does carry its own dangers, however, as genetically identical plants will carry the same susceptibility to disease.

Gardeners can choose a propagation method to suit their needs and the capacity of each plant to reproduce in the local conditions. Knowing the plant family can be a useful guide to rooting cuttings successfully. For example, most plants in the Gesneriaceae, such as African violets (*Saintpaulia*), *Columnea*, *Ramonda* and *Streptocarpus* readily regenerate from leaf tissue. The Lamiaceae, including *Salvia* (sage), *Solenostemon* (coleus), *Lamium* and rosemary, root easily from stem cuttings – in the wild, stems close to moist soil produce roots.

Another factor is the plant's natural limit of distribution; often reproductive ability declines outside this area. This may be countered by providing controlled conditions for propagation.

WHICH PART OF THE PLANT?

Most cuttings are taken from the plant stem, and in many cases several cuttings can be taken from one long length of a plant's new season's growth. They can vary in degree of ripeness from unripe (softwood cuttings) to fully ripe (hardwood cuttings). Cuttings may also be taken from leaves or roots. A few plants can regenerate from a detached leaf or section of leaf tissue. These include members of the families Begoniaceae, Crassulaceae and Gesneriaceae. It is possible to root leaves of plants such as *Clematis*, *Hoya* and *Mahonia*, but leaf cuttings of these cannot produce buds so can never develop into plants. Some plants, such as acanthus, can be propagated from root cuttings.

STEM CUTTING
Stems are used to prepare cuttings of many plants. The current year's growth is selected and it may be soft (as in the hydrangea cuttings shown here), partially ripe or fully ripe.

PART-LEAF CUTTING
A few plants, such as the begonia shown here, regenerate from leaf tissue. Take leaf sections or wound leaves at any time in the growing season.

ROOT CUTTING
Lengths of healthy, strong root can be taken (here from an acanthus) in the dormant season, of pencil or medium thickness for the plant.

Cuttings in the Past

THE CULTIVATION AND PROPAGATION OF PLANTS began when human tribes abandoned their nomadic, hunter-gatherer way of life to live in settled communities. This change occurred just after the last Ice Age and marked the beginning of modern civilization. The development of bread wheat triggered the advent of agricultural practices, and alongside developed techniques for raising plants from seeds and cuttings, as well as grafting rootstocks and layering.

Origins of Propagation

Ancient civilizations throughout the world grew a wide range of food crops, after noting how plants naturally dispersed seeds that later produced seedlings. In ancient Greek and Roman times, writers such as the poet Virgil recorded current methods of propagation in some detail. Olives, date palms and cypresses were grown from seeds as well as other food plants such as cabbages, turnips, lettuces and herbs. Propagation from cuttings began when

> The Romans dipped the bases of cuttings in ox manure

rooted shoots or suckers were detached from the parent plant and replanted, and this led to propagation from unrooted shoots. Romans dipped the bases of cuttings in ox manure to stimulate rooting.

In the Middle Ages, settlers discovered how to propagate superior forms of grapes, olives and figs to preserve their desirable characteristics by thrusting woody stems into the soil. By 2000 BC, grafting was also fairly common in Greece, the Middle East, Egypt and China. Grafting was used to propagate plants that were difficult to root from cuttings, and to encourage early fruiting.

Other natural vegetative reproduction methods were exploited by propagating from food-storage organs such as bulbs,

AARON'S ROD
The Bible story of Aaron's rod, a stave which "brought forth buds, and bloomed blossoms and yielded almonds", is in effect a description of an early hardwood cutting.

tubers and rhizomes. Towards the dawn of the first century AD, plant propagation practices were already well established.

The Victorians

An explosion of plant-hunting took place in the western world in the 18th and 19th centuries. A wealth of new and exciting plants were discovered and traded between Europe and Japan, China, the East Indies, Australasia, Africa, North America, Mexico and South America. New introductions arrived as seeds, bulbs or even plants.

Enthusiasm for these new plants and the desire to grow and propagate them, coupled with the financial wealth of the plant collectors' patrons, was the inspiration for the golden age of the glasshouse. The Victorians were highly inventive in glasshouse construction and design, and their methods of controlling temperature, levels of light and humidity in the glasshouses were impressively complex. Techniques became more refined and propagation techniques increasingly creative, even if trial and error must have been used when attempting to increase stocks of each unfamiliar plant.

Their propagation equipment was fairly basic, yet the ideas of the Victorians form the foundation of techniques used today.

To control temperature and humidity, they built cold frames and hot beds. Cold frames, sited to capture as much warmth as possible from the sun, especially in winter, were used for root cuttings, easy stem cuttings and seeds. Hot beds consisted of a glazed frame set in a pit, filled with equal parts fresh manure and deciduous leaves, a mix that generated heat for up to eight weeks after being mixed and turned.

Bell jars were also used in great numbers. These bell-shaped glass jars were placed over cuttings in prepared soil or in pots, and warmth was provided by solar radiation. Although difficult to control precisely, it was possible to maintain high humidity inside the jars. Today, bell jars have largely been replaced by more versatile cloches.

◀ THE PALM HOUSE AT KEW
Heated glasshouses like the Palm House at the Royal Botanic Gardens Kew enabled Victorian gardeners to grow many tropical plants.

▼ VICTORIAN HOT BED
Victorian gardeners often made use of hot beds for propagation and for raising exotic vegetables or fruits, such as pineapples.

Modern Propagation Techniques

THE PROPAGATION EQUIPMENT that was available to Victorian gardeners was fairly primitive compared with modern advances, yet their ideas still form the basis of what is done today. Since the 1950s, modern technology and increased collaboration and information exchange among professionals has led to the development of new propagation techniques for the first time in centuries. These new methods make propagation much easier today.

Mist and Fog Propagation

A unit known as the intermittent mist propagation system was devised in the 1950s for rooting stem cuttings, in particular softwood and semi-ripe material (*see pp.30–35*). Cables along the base of the unit provide bottom heat, stimulating rooting, and constantly regulated humidity keeps the cuttings cool and moist. A soil thermostat regulates the temperature of the sand or compost bed. The advance meant that many plants that had previously been grafted could now be rooted, at a fraction of the cost. Mist propagation is widely used in commercial propagation and is useful for gardeners. If a dedicated unit is beyond your means, you can create your own using soil-warming cables and a misting system in a closed case.

With fog propagation, a much smaller water droplet is created than with mist propagation, creating conditions in which the air remains moist for a much longer period. The foliage is not misted, making this a good technique for cuttings that are prone to rot.

Plastic Film

Another development of the 1950s was plastic film. An effective propagation environment can be created using a unit with bottom heat with plastic film (a sheet of opaque plastic) fixed over the top.

Use plastic film to warm soil in a cold frame before use

The cover creates a high-humidity environment around the tops of the cuttings. Plastic film can also be used with cold frames to warm soil before cuttings are inserted, and then to cover new plants in the frame.

PROPAGATING BLANKET
This specialist blanket is made of electric wires encased in aluminium foil to provide an even spread of heat. The blanket may be used on a greenhouse bench or on the floor to provide a temporary propagation area, for plant material in pots or in unheated propagators. When not in use, it may be rolled up and stored.

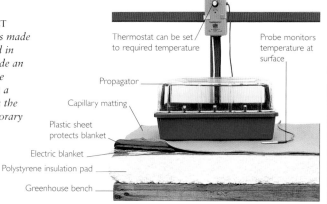

Thermostat can be set to required temperature
Probe monitors temperature at surface
Propagator
Capillary matting
Plastic sheet protects blanket
Electric blanket
Polystyrene insulation pad
Greenhouse bench

MODERN PROPAGATION TECHNIQUES • 13

MIST PROPAGATION UNIT
The misting head sprays fine droplets over the cuttings, cooling the top-growth and retaining moisture. The heated bench aids rooting.

FOG PROPAGATION UNIT
Fresh air is pumped through a water reservoir, creating a warm "fog" around the cuttings. Vapour condenses at the sides and runs back.

MICROPROPAGATION

The technique of micropropagation, used to raise huge numbers of plants from a small amount of material, was developed in the 1960s. New cultivars, virus-free crop plants such as raspberries, and plants that are difficult to propagate by traditional means can be propagated in quantity in this way. It is also a very useful way of increasing old and rare plants from existing stocks, to conserve plants in the wild. This is a commercial technique only, but the end results are of benefit to gardeners.

Micropropagation usually involves growing pieces of plant tissue *in vitro* in sterile laboratory conditions. Most plants are able to regenerate from a single cell. Tissue from the shoot tip (meristem) is most commonly used, but root tips, calluses (which form on wounds), anthers, flower buds, leaves, seeds or fruits may also provide suitable tissue. In specially adapted growing rooms, temperature, light levels, nutrients and hormones are regulated. The resulting plants are then grown on in greenhouse conditions.

Viruses and systemic disease rarely penetrate growing tips, so micropropagated plants are normally disease-free and therefore may be safely introduced to other countries.

Although micropropagation is capable of producing large quantities of plants, it is still an expensive procedure. Another disadvantage is that plants may fail to adapt to a normal growing environment.

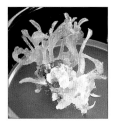

CULTURED PLANT TISSUE
Plant cells (here of tobacco plant, Nicotiana) are grown on a nutrient gel until the cell mass produces embryo plants.

LEAF CUTTING
Plantlets can be grown from a tiny leaf cutting, as with these African violets. In sterile conditions, better yields can be obtained.

REGENERATION

THE PROCESS OF TAKING CUTTINGS is relatively simple in most cases, but success will depend on several factors. The inherent ability of the parent plant to produce roots will determine the degree of care needed to coax cuttings to root. Also, the condition of the parent influences the quality of the rooted cutting. Always choose a healthy plant; diseases or pests can be transmitted to a cutting. Good hygiene is essential to avoid introducing disease.

SUCCESS WITH CUTTINGS

Plants have a remarkable ability to regenerate themselves from little more than a piece of the stem, leaf, root or a bud, into a fully developed plant with a root system and shoots. Vegetative propagation exploits this natural ability and extends it.

To produce these roots, a group of growth cells (known as the meristem), which are usually close to the central core of sap-carrying, or vascular, tissue, change into root cells. These then form root buds and then adventitious roots. These are also called "induced" or "wound" roots because, in most plants, they occur only after wounding, such as when a piece of stem is cut off to produce a cutting.

In some plants, preformed root initials lie dormant in the stems and so they root rapidly and easily from cuttings. A few plants even form root buds, normally visible at the bases of shoots. Other, often hardy, woody plants are difficult to root; with these, callusing may hinder root formation and grafting may be necessary. In warm climates, cuttings of many plants may be rooted outdoors, directly inserted into prepared soil in shade, at almost any

◄ CLINGING IVY
Ivies (Hedera) develop stem roots, known as adventitious roots, that cling to vertical surfaces.

▲ IVY ROOTS
The stem roots are clearly visible against the wall. These evergreen climbers root readily in the wild.

time of year. In cooler areas, a controlled environment is often vital; rooting may be unpredictable and slow. Bottom heat of 15–25°C (59–77°F) can promote rooting. The air should be much cooler than the growing medium, to avoid encouraging growth of foliage instead of roots. The rooting medium should be moist and (for leafy cuttings) the air kept humid.

The time taken for a cutting to root depends on the plant, the type of cutting, the age of the stem, how it was prepared

> A piece of leaf may have the ability to grow into a fully developed plant

and the rooting environment. Leafy cuttings, particularly softwood and greenwood cuttings of woody plants and perennials (including basal cuttings) may root in about three weeks; woody cuttings from trees and shrubs may take up to five months. The new plants are genetically identical to the parent, and characteristics such as variegated leaves can be preserved.

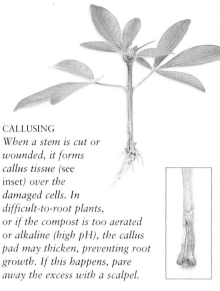

CALLUSING
When a stem is cut or wounded, it forms callus tissue (see inset) over the damaged cells. In difficult-to-root plants, or if the compost is too aerated or alkaline (high pH), the callus pad may thicken, preventing root growth. If this happens, pare away the excess with a scalpel.

NODAL CUTTING
The cells involved in growth are most concentrated at the nodes, or leaf joints, so most cuttings are trimmed just below a node to optimize root formation.

Use a clean, sharp knife

WOUNDING
A cutting from semi-ripe or hard wood often roots more readily if bark is cut away from the base of the stem. This exposes more of the growth cells in the cambium layer.

2.5cm (1in) cut exposes growth cells

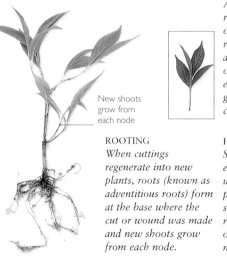

New shoots grow from each node

ROOTING
When cuttings regenerate into new plants, roots (known as adventitious roots) form at the base where the cut or wound was made and new shoots grow from each node.

HEEL CUTTING
Some cuttings, especially of semi-ripe wood, are taken by pulling away a small sideshoot so that it retains a "heel" of bark from the main shoot.

Heel contains high levels of growth hormone

WOODY PLANTS

TREES, SHRUBS, CONIFERS AND WOODY CLIMBERS form the backbone of any garden planting, but vary enormously in habit, form and productive lifespan. They can be propagated by an equally wide range of techniques, but taking cuttings is one of the most common propagation methods, especially where a large number of new plants is required (for hedging for example). It is usually a fairly simple process and provides new plants relatively quickly.

SHRUBS AND WOODY CLIMBERS

Shrubs and climbing plants represent an invaluable and long-lasting source of shape, texture and colour in the garden. A shrub is a deciduous or evergreen perennial with multiple woody stems or branches, generally originating from or near its base. Subshrubs are woody-based plants with soft-wooded stems. Climbers are plants that climb or cling by means of modified stems,

> The majority of shrubs and climbers are best propagated by cuttings

roots, leaves or leaf-stalks, using other plants or objects as support.

Choosing the type of cutting, and the ripeness of the wood, is very important to the success of the propagation process.

It is important to select cutting material very carefully, avoiding any shoots where pests or diseases may be present, and discarding any damaged material. Cuttings root most easily when the parent plant is young and producing good lengths of new growth each year. Most cuttings will be taken from the current season's growth.

HONEYSUCKLE
Climbing honeysuckles like this Lonicera periclymenum *'Serotina' are easily propagated from different types of growth: softwood, semi-ripe or hardwood cuttings, taken from late spring to winter.*

The cuttings of some shrubs, such as deciduous azaleas and magnolias, root best if the material is forced into early growth under protection early in the year.

GARDEN TREES

Trees may provide the framework or focal point of a garden, and can also link the garden with the landscape beyond. They are woody perennials with a crown of branches, usually at the top of a single stem or trunk and include conifers, or cone-bearing trees. Valued for their shape, which

provides year-round interest, many trees also offer seasonal displays of handsome foliage as well as bark, flowers and brilliant berries. Since trees are slow-growing compared with herbaceous plants, they can be expensive, so it is worth propagating your own, especially if a number of plants are needed for hedges, woodland gardens or screening. Propagating also makes it possible to increase unusual species, or to replace old trees.

Most hardwood cuttings will yield a sapling ready for planting out in the open in one year; other types of cutting will need to be grown on for two to three years. However a few plants, *Nothofagus* for example, can take up to five years.

▶ HOLLY
Variegated hollies like this example of Ilex aquifolium *'Ovata Aurea' keep their colouring when grown from cuttings. Semi-ripe cuttings of this ready-rooting holly can be taken fom late summer onwards.*

▲ CAMELLIA
This Camellia × williamsii *'E.G. Waterhouse' could be increased by taking semi-ripe cuttings from midsummer on.*

▶ CONIFER
Chamaecyparis lawsoniana *'Minima Aurea' can be grown from semi-ripe cuttings taken in late summer or autumn.*

PERENNIALS

THIS HUGELY VARIED GROUP OF PLANTS encompasses not only traditional border herbaceous perennials but also alpines, water garden plants, orchids and tender plants for summer bedding and greenhouse display. Many perennials are easily propagated from cuttings, enabling the gardener to keep existing plants healthy and vigorous, replace short-lived perennials as they fade, and build up stocks for an effective border display.

MAINTAINING THE LINE

The majority of perennials make new growth from the base or crown; their roots or rhizomes spread and the plants naturally form clumps. To give impact to plantings, perennials are often required in quantity – cuttings provide the means. They also provide the best way of obtaining plants that are true to the parent, with all their special characteristics such as blooms of a particular colour, or large or double flowers. This also applies to plants bred not to flower such as the lawn chamomile 'Treneague'; foliage plants with finely cut, differently coloured or variegated leaves; single-sex plants; and sterile hybrids.

TYPES OF CUTTINGS

A wide range of perennials can be raised from cuttings, using a variety of plant parts – stems, leaves and roots. In most cases, some form of controlled environment – a heated propagator, greenhouse, or cold frame – is necessary to encourage cuttings to regenerate "missing" parts such as roots. If these conditions can be provided, cuttings are ideal for obtaining a number of new perennials that will be ready to plant out, and may even flower, in their second year.

Mature plants recover well from having a modest amount of cutting material removed. Stock plants can be cultivated especially for the purpose of providing cuttings. They should be compact and bushy, with lots of young shoots.

TIMING

With some perennials, you can take cuttings at almost any time of the year they are not in flower. With others, material is only suitable during a few weeks or even days. If taken after flowering, many cuttings will root and grow well. Cuttings from perennials that die down over winter should be taken early in the growing season to give them a long growing period.

OENOTHERA MACROCARPA
Ozark sundrops may be easily propagated from softwood tip cuttings or basal stem cuttings taken in the spring.

▲ ANEMONE
To avoid disturbing Japanese anemones like this Anemone × hybrida *'Max Vogel', uncover the edge of the clump and take root cuttings.*

▶ PINKS
*Semi-ripe cuttings, or "pipings", may be taken from all pinks (*Dianthus*), especially small and alpine species. Plants should flower the next year.*

▲ MINT
Take rhizome cuttings of mint (here Mentha suaveolens *'Variegata') in spring and root with bottom heat of 10°C (50°F).*

▼ DRUMSTICK PRIMULA
Root cuttings can be used to propagate Primula denticulata *and its colour forms. Cut thicker roots of the parent plant into pieces (see p.44).*

Cacti and other Succulents

SUCCULENTS EVOLVED TO SURVIVE in habitats with extreme conditions, particularly periods of drought. The sculptural forms of this extraordinary group of plants belie the comparative ease with which many of those in popular cultivation may be propagated. Cacti form one group of stem succulents, distinguished by the areole, a pad-like bud from which flowers, shoots and spines grow. Other succulents span many plant families.

Diversity of Forms

Succulents store water in specialized tissue in swollen roots, stems or leaves. Many desert species have tiny leaves, or no leaves at all, to retain moisture and withstand drought. Others are rainforest epiphytes, living in trees and absorbing water through strap-like stems. Other succulents are very diverse in form, from stark, cactus-like barrels to tree-like leafy species, and are also diverse in the ways they may be propagated. Some techniques, such as stem and leaf cuttings, are broadly similar to those used on herbaceous perennials, but with the advantage that succulent cuttings do not wilt as quickly. However, fleshy cuttings are very susceptible to rot, so good hygiene is essential for successful propagation of these plants.

Some cacti and other succulents do not flower readily in cultivation, and commercial seeds are often not readily available, so taking cuttings offers a reliable way of increasing many of these plants. Succulent cuttings have the advantage that, because of their fleshy tissue, they can retain nutrients and water while they become established.

Unusual forms, such as variegated, monstrose or cristate (crested) plants, and hybrids, can usually only be propagated from cuttings (or by grafting) to preserve their distinctive characteristics.

Types of Cutting

There are various types of cuttings, and the most suitable technique depends on the plant's form and growth habit. Succulents are generally propagated by stem, leaf, or rosette cuttings, while cacti are raised from globular, columnar or flat stem cuttings. Many clump-forming species produce unrooted offsets, which may also be treated as cuttings.

The best time to take cuttings of most succulents, especially in cool climates, is in late spring or early summer when the warmer, drier weather arrives and the plants have started to grow strongly. It then gives them a chance to establish for as long as possible before the following winter.

CRASSULA
In the wild, crassula leaves that have fallen on the ground often take root. Crassula arborescens, the silver jade plant, shown here, can be rooted from semi-ripe stem cuttings.

CACTI AND OTHER SUCCULENTS • 21

▲ SUCCULENTS
Cuttings offer a reliable way of increasing succulents, as they retain water while rooting.

▶ CACTI
Offsets taken from cacti like Mammillaria zeilmanniana *may be treated in the same way as stem cuttings.*

◀ ORCHID CACTUS
The orchid cactus, Epiphyllum crenatum, *can be easily propagated from flat stem cuttings taken from spring to late summer. The cuttings should root in 3–6 weeks.*

RAISING PLANTS FROM CUTTINGS

CHOOSING AND COLLECTING

THE FIRST PRINCIPLE OF PROPAGATION is to take material from healthy, strong plants, because pests and diseases can be transmitted from the parent. Look for clean, vigorous growth, avoiding weak or very spindly shoots, and make sure that your standards of hygiene are high. Choosing the type of cutting, and the maturity of the material, best suited to a particular plant, is very important to the success of the propagation process.

UNDERSTANDING THE BASICS

Cuttings from trees, shrubs and woody climbers root most easily when the parent plant is young and producing good lengths of new growth each year. Most cuttings will be from wood of the current season's growth. Some shrubs, such as deciduous azaleas and magnolias, generally root best if the material is forced under protection early in the year.

Stem cuttings from perennials that die down over winter should be taken early on in the growing season, so that the cuttings have plenty of time to form good root systems. They will need these to come through the next dormant phase.

Hardwood cuttings from woody plants should be taken during the dormant period, usually from mid- to late autumn through to late winter. Softwood cuttings are usually taken in late spring from the fast-growing tips of new shoots. Semi-ripe cuttings are taken during the summer.

Always use very clean, sharp cutting tools and never allow cuttings to dry out, either when preparing them or when growing on.

CUTTINGS FROM STOCK PLANTS

A stock plant is grown purely to provide cutting material and can be encouraged to produce the best type of growth for this purpose. A stock plant should be healthy, mature and vigorous, with compact, bushy growth and lots of young shoots. It should be a good example of its type, and flower and fruit well. No more than 60 per cent of the top-growth should be taken from a stock plant at any one time.

REGULAR SUPPLY
A container-grown plant can supply cuttings repeatedly. This hebe yielded 84 cuttings without its shape being spoiled.

◀ PLANTS TO PROPAGATE *Honeysuckle, broom and hebes are all easily raised from cuttings*

Tools and Sundries

An assortment of items will prove either useful or essential in the preparation of propagating material, from basics such as garden knives and pots to specialist propagators. A small, but essential, item is the label: always label propagation trays to avoid confusion later. When preparing to take cuttings, it is important to use the most appropriate knife. Dibbers are handy for making planting holes, and widgers are useful for lifting delicate cuttings.

Knives and Cutters

Use a garden or budding knife with a plastic or wooden handle for softwood or semi-ripe cuttings. Most have a carbon steel blade, which folds into the handle. Use a scalpel or fine-bladed craft knife for very small cuttings and for cutting soft tissue such as cacti. Secateurs are good for woody cuttings.

◀ SHARPENING STONE
Use this to keep the blades of knives and secateurs sharp. A sharp blade will not crush delicate plant tissue.

▲ SECATEURS
Secateurs are used to prepare woody material, such as hardwood cuttings. By-pass secateurs, shown here, are better than anvil types as they do not bruise the stems.

LABELLING
Plastic labels (below) are widely used when propagating plants from cuttings. They may be written on in pencil, so are reusable, although writing does fade and the plastic becomes brittle over time.

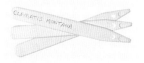

| GARDEN KNIFE | BUDDING KNIFE | SCALPEL | STEEL WIDGER | PLASTIC WIDGER | SMALL DIBBER |

Plant Pots

A wide range of containers, including the traditional pot and seed tray, is available for propagation purposes. Plastic is more hygienic, lighter and cheaper than clay or terracotta. Plastic pots also retain more moisture, although clay pots offer better aeration and drainage and are a good choice for alpines. Flexible plastic and soft plastic pots are cheaper than rigid pots, but are generally used only once.

POTS FOR PROPAGATION
An assortment of sizes will be needed. The most useful are the 9cm (3½in) and 12.5cm (5in).

MODULE TRAYS

Module, or cell, trays have been used commercially for a number of years and are now widely available to the amateur gardener. The modules allow cuttings to develop sturdy root systems before being potted up. The modules also mean the young plants can be handled without disturbing roots or harming the stems. Care is needed in watering, because the trays can dry out quickly. Trays in a range of sizes (in cells of 13mm, 20mm, 30mm and 37mm) are now available for propagation purposes; a 37mm module tray is the most appropriate size for small herbaceous cuttings. Trays of modules of rockwool (a porous material made from fibres spun from molten mineral rock) can also be used, but the cuttings will need to be fed with a dilute liquid fertilizer as soon as they have rooted. Standard or half seed trays may also be used for rooting all kinds of cuttings.

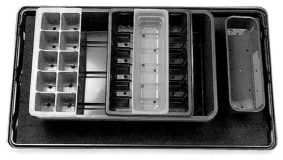

▲ TRAYS AND INSERTS
Trays made of plastic fit into a standard seed tray. Drip trays with capillary matting allow watering from below.

▶ MODULE TRAYS
These module or cell trays in a range of sizes allow rooted cuttings to be potted without too much root disturbance.

THE IMPORTANCE OF HYGIENE

It is essential to maintain high standards of hygiene when propagating plants, in order to prevent any possible contamination by pests and diseases. Sterilize tools and equipment before use, either by heating them or wiping them in surgical spirit before each cut. Dirty containers can harbour minute pests; make sure they are throughly scrubbed, rinsed and dried before use. Close-fitting latex gloves are ideal for use when propagating. They are sterile and also protect your skin against irritant sap.

RUBBER GLOVES

STERILIZING KNIVES
To sterilize a knife, dip the blade in methylated spirits and pass it quickly through a flame.

WASHING POTS
Thoroughly scrub dirty pots using a stiff brush dipped in dilute horticultural disinfectant. Rinse well.

Rooting Media

THE RIGHT ROOTING MEDIUM is crucial to success with cuttings. Soil beds outdoors are often used for hardwood cuttings, but most cuttings need composts and other media under cover. You can mix your own compost to obtain the ideal medium for individual plants. Most cuttings composts are based on peat or peat substitutes, combined with other ingredients with different properties. Sterile, inert substances such as perlite and vermiculite are useful.

Common Ingredients for Compost

When rooting cuttings under cover, compost is usually preferred to soil, because it is light, well-aerated and relatively free from pests and diseases. Compost mixes intended for rooting cuttings need to be free-draining. A standard cuttings compost typically contains equal parts of sand and peat substitute (or peat). It may also be based on bark or perlite or a high proportion of coarse sand (river sand). Composts like these are low in nutrients so the cuttings will need feeding once rooted. Alternatively, for cuttings that will be in the pot for some time, such as those for woody plants, add a little fertilizer – to the bottom of the pot so that the new roots will not be scorched.

LOAM
High-quality, sterilized soil, with good nutrient supply, drainage, aeration and moisture retention. For soil-based composts.

GRIT
Used in 2–3mm very fine (right) or 5mm fine (left) to 7–12mm coarse grades. Aids drainage for alpine and cactus composts.

PEAT
Stable, long-lasting, well-aerated and moisture-retentive, but low in nutrients. For lightweight, short-term mixes.

PERLITE
Expanded volcanic rock granules. Sterile, inert and light, retains moisture but drains freely. Medium/coarse grades aid aeration.

FINE BARK
Fine grades of chipped pine bark used as a peat substitute, or for very free-draining, acidic composts, good for cuttings composts.

VERMICULITE
Expanded and air-blown mica, acts in a similar way to perlite but holds more water and less air. Fine grade aids drainage.

COIR
Fibre from coconut husks, used as peat substitute. Dries out less quickly than peat, but needs feeding. Base for soilless mix.

SAND
Fine (silver) sand (left) helps drainage and aeration; coarse sand (right) gives more open texture to cuttings compost.

LEAF MOULD
Well-rotted, sieved leaves, used as a peat substitute. A non-sterile medium, leaf mould may harbour pests or diseases.

GARDEN SOIL

If you are planning to use garden soil in home-made compost mixes, it must first be sterilized to kill off harmful organisms. To do this, the soil must be sieved to remove stones, then heated. Use a conventional oven (30 mins at 200°C (400°F) or microwave in a pierced roasting bag on full power for ten minutes.

COMPRESSED PEAT BLOCKS

These small, biodegradable blocks of peat or coir contain a special fertilizer. The compressed blocks more than double in size when soaked in water for 10–20 minutes. A plastic mesh holds the peat together. Once soaked, they form individual planting modules and cuttings can be inserted into the hollow at the top of each block. Do not let blocks dry out, and pot or plant cuttings, complete with blocks, as soon as roots show.

SOAKED

UNSOAKED

INERT GROWING MEDIA

Different kinds of sterile, inert media can also be used for cuttings, avoiding the problem of harbouring pests and diseases associated with soils and composts. Damping off is also less of a problem with sterile media. The plants basically grow in water, with nutrients added directly to the water in the form of liquid fertilizer. The plants also have access to unlimited oxygen, as the plant roots are in direct contact with the air. Porous rockwool (fibres spun from mineral rock) is used in commercial trade, but other options for the gardener are florist's foam, perlite, gel, sand, pumice or grit.

ROCKWOOL LOOSE FIBRES

ROCKWOOL LOOSE GREENMIX

ROCKWOOL MODULES, OR "CUBES"

▶ FLORIST'S FOAM
Water-retentive and with a light, open texture, florist's foam can be used for soft cuttings such as fuchsias.

WATER-RETENTIVE GEL AND WATER

Water-retentive gel is commonly used in container composts to conserve water. The dry crystals absorb water, and increase in volume, forming a granular jelly. The gel can be used on its own for rooting woody cuttings, such as yew (*Taxus*); add a liquid fertilizer to the water used to hydrate the gel, insert the cuttings and keep sealed until they root. Stem-tip cuttings of easy-to-root perennials, for example penstemons, *Solenostemon* (coleus), and *Impatiens*, may be rooted in water. Place the cuttings in a jar of water on a greenhouse bench or windowsill.

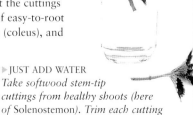

GEL BEFORE SOAKING

AFTER SOAKING

▶ JUST ADD WATER
Take softwood stem-tip cuttings from healthy shoots (here of Solenostemon). Trim each cutting just below a leaf joint; remove the lower leaves. Place netting over the jar of water to hold the cuttings in place.

Preparing Cuttings

Raising new plants from cuttings is frequently a very straightforward process, and it is certainly the most widely used technique for propagating the majority of shrubs and climbers. Collect material early in the day before the sun and rising temperatures cause plants to lose moisture through their leaves. You can either prepare the cuttings immediately or store them in plastic bags in a cool place (such as a fridge), out of direct sunlight, for a few hours.

How Shoots Ripen

Cuttings can be taken from different parts of a plant shoot. This shoot from a firethorn (*Pyracantha*) demonstrates the different stages of woodiness seen in a stem as a plant grows and develops. The softwood at the tip is still green, soft and sappy, while the greenwood in the middle is less flexible. The base of the shoot is semi-ripe, becoming woody and dark.

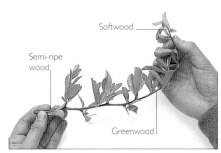

Trimming a Cutting

Most cuttings are taken from a plant stem; they may be severed between the leaf joints or nodes (internodal cutting) or just below a node (nodal cutting). The growing tip may also be removed to redistribute natural growth hormones to the rest of the stem.

SELECTING THE TRIMMING POINT
Cuttings are usually trimmed just below a leaf joint or node, where the growth hormones accumulate (see near right). Easily rooted plants can be cut between nodes (see far right).

NODAL CUTTING   INTERNODAL CUTTING

Using Fungicide and Rooting Powder

Cut large leaves in half to reduce moisture loss and immerse very soft cuttings in fungicidal solution. Dip the base into hormone rooting compound or gel, which will encourage root formation, and insert each cutting in the compost, just deep enough for it to be able to stand upright.

FUNDAMENTAL HYGIENE
It is a good idea to wear close-fitting, sterile gloves when preparing cuttings, as it helps to maintain a hygienic environment. If you have no gloves, make sure your hands are clean.

PREPARING CUTTINGS • 29

DIFFERENT TYPES OF CUTTING

Cuttings are taken from stems, leaves or roots and there are various types, depending on the plant. In some plants, such as ivy (*Hedera*), and many in the mint family (including rosemary and salvias), pre-formed embryo roots lie dormant in stems and so they root rapidly from cuttings. Other, often hardy, woody plants may be difficult to root. It is important to choose the right type of cutting.

SOFTWOOD CUTTING
Cotoneaster

HEEL CUTTING
Ceanothus

LEAF-BUD CUTTING
Clematis

ROSETTE CUTTING
Saxifraga

LEAF CUTTING
Sedum

HARDWOOD CUTTING
Salix

STEM-TIP CUTTING
Fatsia

SEMI-RIPE CUTTING
Abelia

STEM CUTTING
Yucca

LEAF-BUD CUTTINGS
Mahonia

CHEVRON LEAF CUTTING
Streptocarpus

LEAF CUTTING
Sansevieria

BASAL STEM CUTTING
Delphinium

ROOT CUTTING
Ailanthus

Soft- and Greenwood Cuttings

Softwood cuttings can be collected from the plant from spring to early summer, before the new growth has begun to firm. This method is suitable for most deciduous shrubs and climbers, and can be used for various deciduous trees including cherries. Softwood cuttings are usually taken from fast-growing tips of new shoots and typically root very easily. Greenwood cuttings are similar to softwood, but are taken when the new growth is just beginning to firm.

Choosing Softwood Cuttings

Cuttings from shrubs and climbers root most easily when the parent plant is young and producing long lengths of new growth each year. Most cuttings will be taken from wood of the current season's growth. Alternatively, use plants bought from the local garden centre, which will invariably have been grown under protection, as stock plants. Increase the yield of cuttings by taking one stem-tip cutting and several stem cuttings from the same stem. The best material is usually the new growth that is neither very thin and weak, nor very vigorous; the latter is often hollow and prone to rotting. Choose instead the material in between these two extremes, which normally has quite short internodal growth (the distance between two sets of leaves). This technique is also suitable for various trees, including ornamental cherries (*Prunus*), some maples (*Acer*), *Liquidambar*, birches (*Betula*) and elms (*Ulmus*).

SOFTWOOD WALLFLOWER CUTTING

Greenwood Cuttings

Greenwood cuttings can be dealt with in a similar way to softwood, but are taken when the new growth is just starting to firm, between late spring and midsummer. Greenwood material is easier to handle than softwood, because it doesn't wilt so readily, although it is treated in the same way. Should you miss the softwood season, greenwood cuttings of most deciduous plants and some evergreens generally root just as well as softwoods.

SOFTWOOD CUTTING
Select sturdy, healthy new shoots in spring or early summer for softwood cuttings (here of a rose species), before they start to firm up or become woody.

GREENWOOD CUTTING
Take greenwood cuttings from vigorous current year's shoots (here of Philadelphus*) that are firm and slightly woody at the base. Prepare as for softwood cuttings.*

GREENWOOD CUTTINGS OF LIQUIDAMBAR
These cuttings have been trimmed at the tip above a node and stems have been wounded. Leaves are halved to reduce moisture loss.

Taking Softwood Cuttings

Softwood cuttings should usually be 4–5cm (1½–2in) long, with one to three pairs of leaves retained at the top. Collect material early in the day before temperatures begin to rise. Remove the soft tip from each cutting because it is vulnerable to both rotting and scorch. This also ensures that once rooted, the cutting does not immediately grow upwards from the tip alone, and ensures a bushy plant from the start. Remove the lowest pair of leaves to make it easier to insert the cutting into the compost. On delicate material this should be done cleanly with a sharp knife or secateurs; where there is no risk of damaging the stem, pinch off the foliage between thumb and forefinger. Dip the base in hormone rooting compound. With softwood cuttings, it is best to make a hole in the compost with a dibber or pencil.

Soft tip removed from shoot

Cut large leaves in half to reduce moisture loss

Prepared cutting

Lower pair of leaves removed

1 In early spring to early summer, cut off non-flowering, vigorous shoots (here *Hydrangea macrophylla*) with 2–3 pairs of leaves. Use secateurs to cut a stem just below a pair of leaves.

2 To prepare each cutting, remove the soft tip from the shoot just above a leaf and then remove the lowest pair of leaves. The stem of the cutting should be about 4–5cm (1½–2in) in length. Large leaves may be cut in half to reduce moisture loss.

Propagator vent

3 Fill 13cm (5in) pots with standard cuttings compost and space the cuttings around the edge. The leaves should be just above the compost surface and should be placed so that they are not touching.

4 Water the cuttings with a fungicidal solution, label, and place the pots in a propagator. Leave in a shaded place. Provide bottom heat of 15°C (59°F) to encourage the rooting process.

5 Once the cuttings have successfully rooted, wean them off. Gently tease apart and pot individually into 9cm (3½in) pots. Cuttings that were pinched out will produce bushy growth.

Semi-ripe Cuttings

THIS TYPE OF CUTTING involves shoots of the current season's growth that have begun to firm up. The base of the cutting should be quite hard, while the tip of the cutting should still be actively growing and therefore still soft. This propagation method is suitable for a very wide range of woody plants, including both evergreen and deciduous species. Semi-ripe cuttings are useful for producing large numbers of plants, for example for hedging purposes.

Taking Semi-ripe Cuttings

The best time to take semi-ripe cuttings is from mid- to late summer, or even in early autumn. In warm climates, growth may be semi-ripe in early summer. The length of the cutting is dependent on the habit of the plant, but between 6–10cm (2½–4in) is suitable for most shrubs and climbers, 10–15cm (4–6in) for broad-leaved evergreens such as magnolias, laurels and hollies. Choose a healthy stem and remove any sideshoots to prevent moisture loss in the cutting.

1 In mid- to late summer, select a healthy shoot of the current season's growth (here from a spotted laurel, *Aucuba*). Take cuttings early in the day, and use clean, sharp secateurs to sever the cutting just above a node.

2 Put the shoot in a clear plastic bag and label it if it is not to be prepared immediately. If necessary, store in a cool place out of direct sunlight for a couple of hours (at most) or place in a refrigerator for a few days.

3 Remove the sideshoots from the main stem. Trim the cutting to 10–15cm (4–6in) long, cutting just below a node. Remove the lowest pair of leaves and the soft tip.

4 Make a shallow wound on one side of the stem by carefully cutting away a piece of bark 1–2cm (½–¾in) long from the base of the stem. This will help to stimulate rooting.

5 Dip the base of the cutting, including the entire wound, into some hormone rooting compound (here in powder form). Root with or without bottom heat according to subject.

Semi-ripe Cuttings of Camellia

Most of the fully hardy to frost-tender, evergreen camellias root from semi-ripe cuttings. They need care and free-draining compost in cool climates, but are easy to propagate in warmer regions. Take cuttings from midsummer to early autumn. Cuttings may be internodal or nodal (see below for variations), with 1.5cm (⅝in) wounds, but nodal tip cuttings produce a flowering plant quickly, in 3–4 years. Apply hormone rooting compound sparingly on nodal leaf-bud cuttings. Root the cuttings in humid conditions and provide bottom heat of 20°C (68°F).

| NODAL LEAF-BUD | INTERNODAL LEAF-BUD | NODAL STEM-TIP | SPLIT LEAF-BUD | THREE-NODED STEM |

Taking Heather Cuttings

There are three principal heaths and heathers: *Calluna*, *Daboecia* and *Erica*, and propagation is always by cuttings (or layering), because seeds do not come true. Of all the heathers, cuttings from *Daboecia* and *Erica* root most readily and are least prone to disease. Take semi-ripe cuttings from healthy, vigorous, non-flowering shoots. Remove the lower leaves from *Erica* and *Daboecia* as a precaution against rot. Use of hormone rooting compound is unnecessary, and nitrogenous fertilizer should be avoided as heaths and heathers are sensitive to the salts. Some varieties root faster than others, so insert the cuttings individually in modules. For best results, root the cuttings in a heated propagator at 15–21°C (59–70°F). Spray or water in a general fungicide and remember to ventilate the cuttings daily.

1. Trim each stem (here of *Calluna*) to a length of about 4–5cm (1½–2in). Hold the base of the cutting firm, and cut straight across the stem at the appropriate point with a clean, sharp knife.

2. Stripping lower leaves from callunas is optional; lightly pinch each stem about one third from the base and quickly pull it through finger and thumb. Pinch out the tips of all cuttings.

3. Use a mixture of equal parts moist coir and peat, or equal parts fine bark and peat. Insert cuttings so that the lowest leaves just rest on the compost surface. Do not firm in the cuttings.

Taking Conifer Cuttings

Most conifers can be raised in a variety of ways, but taking semi-ripe cuttings is the easiest for many types. Conifers are usually propagated from the current year's growth and the basic principles are the same as for other trees and shrubs, except that conifers make new growth in two ways: main or leading shoots grow straight upwards and side shoots grow outwards. With most conifers it is very difficult to make a cutting taken from a sideshoot form a main stem (although with pines and cypresses, there is no problem). Therefore the best cutting material is from strong leading shoots. Cuttings taken from young (juvenile) growth usually root best.

Ensure foliage sits just above compost

1 Take neat young shoots, not adult ones with fruits. If needed, strip off the sideshoots or needles from the bottom third of each stem (here of *Chamaecyparis* 'Chilworth Silver'). The small wounds promote rooting.

2 Dip the base of each cutting in hormone rooting compound (here powder). Insert easily rooting cuttings singly in 8cm (3in) pots; make a hole, insert a cutting, firm and water. Spray with a fungicide to prevent rot.

Pipings from Pinks and Carnations

Semi-ripe cuttings, known as "pipings", can be taken from all kinds of pinks and carnations (*Dianthus*), especially small and alpine species. Take cuttings from mid- to late summer. Once the cuttings have been removed from the plant (see right), dip the bases into hormone rooting compound, and insert the stems into pots of cuttings compost. Place the pots in a frame or propagator and keep moist but not wet. Rooting takes 2–3 weeks at 15°C (59°F), and plants should flower the following year.

TAKING PIPINGS FROM PINKS
Choose a healthy plant first. Hold a non-flowering shoot near the base and pull out the tip. It should break easily at a node, yielding a cutting 8–10cm (3–4in) long with 3–4 pairs of leaves. Remove the lowest pair (see inset).

Taking Heel Cuttings

For plants that are difficult to root, it is a good idea to take heel cuttings, where a sideshoot is pulled from the plant so that it retains a "heel" of bark. The heel forms an area where the natural rooting hormones of the plant build up, improving chances of success in rooting the cutting. It also provides a hard end-point to the cutting, which is less prone to fungal attack. Many *Ceanothus* species are rooted in this way.

BEFORE

Trim off "tail" just below node

AFTER

1 Carefully pull away a healthy sideshoot of the current season's growth (here of a *Ceanothus*), and make sure that it comes away with a sliver, or "heel", of bark from the parent shoot. The sideshoot should be about 10cm (4in) long.

2 Trim off the "tail" of the heel with a clean, sharp knife, holding the cutting firmly. The heel contains growth hormones that will encourage rooting. Rooting is as for ordinary semi-ripe cuttings, with or without bottom heat, depending on the subject.

Taking Mallet Cuttings

You can propagate *Berberis* from semi-ripe cuttings, but some species and cultivars, like *Berberis* × *lologensis*, root better from mallet cuttings. These are shoots of current growth with a piece of wood from the previous season forming a mallet-shaped plug at the base. Apply hormone rooting compound for best results.

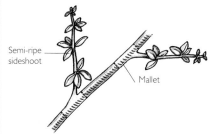

Semi-ripe sideshoot

Mallet

Mallet

1 Take mallet cuttings from last year's stems, in early summer for deciduous species or autumn for evergreen species. Choose short sideshoots, about 10cm (4in) long, of the current season's growth. Cut just above and below the joint on the main stem to leave a 1cm (½in) section (or "mallet") at the base of each cutting.

2 Remove the lower leaves and soft tip of each sideshoot. Slit the mallet lengthways if its diameter is more than 5mm (¼in). Then treat the cuttings as semi-ripe cuttings. This method gives thin-stemmed cuttings a more substantial base from which to produce roots. Mallet cuttings are often used to propagate shrubs with pithy or hollow stems.

Hardwood Cuttings

MANY SHRUBS AND CLIMBERS can be propagated from hardwood cuttings, including dogwoods, willows, and honeysuckles. It is also one of the easiest and cheapest ways of raising many deciduous trees. Bush fruits, especially blackcurrants, redcurrants and gooseberries are also increased in this way. Long lengths of fully mature, young stems are taken, after leaf fall and before new spring growth, from deciduous, woody plants or broad-leaved evergreens.

Taking Hardwood Cuttings

Deciduous subjects are propagated from late autumn to midwinter, once the current season's growth has completely matured. Evergreen cuttings are taken at a similar time, when the leading growth bud is resting and the new growth has fully matured. Hardwood cuttings are normally much bigger than softwood or semi-ripe ones, since they are much slower to root and need extra food reserves for winter.

1 From late autumn, take well-ripened shoots of deciduous shrubs, trees or climbers (here *Forsythia*). Remove any remaining leaves.

2 Cut off each shoot at the base of the current season's growth. Trim off the tip of each shoot if it has not ripened. Cut the shoots into 20cm (8in) sections (about the length of a pair of secateurs). Make a horizontal cut just below a node at the base and a cut sloping away from a bud at the top.

3 Prepare a slit trench in free-draining soil: push the spade into the soil about 15cm (6in) down and press forward. Dip the base of each cutting into rooting compound (*see inset*).

4 Insert the cuttings about 5cm (2in) apart so that about a quarter of each is visible. Rows of cuttings shuld be 30cm (12in) apart. Backfill the trench and firm the soil. Water in.

Slow-rooting Hardwood Cuttings

For tree species that do not root easily, such as dawn redwoods (*Metasequoia*), or laburnums, overwinter bundles of cuttings in sand (*see right*). Each bundle should have no more than ten cuttings, otherwise the middle ones will dry out. Sand will allow the cuttings to undergo a period of cold but will protect them against wide fluctuations in temperature. Use washed sharp sand with a low clay content so that the surface does not form a crust. Make sure it is moist, especially after periods of frost which can dry it out. Leave the cuttings until just before bud break in early spring, then line them out in a nursery bed.

1 Tie the cuttings (here *Metasequoia*) into bundles of up to ten cuttings. Dip into hormone rooting powder or gel.

2 Insert the bundles into a bed of sand in a sheltered place or cold frame over winter. They should root by spring.

Extra-long Hardwood Cuttings

The simplest way of propagating most fast-growing trees like poplar and willow is to take extra-long hardwood cuttings. Take the cuttings after leaf fall and select vigorous, straight shoots up to 2m (6ft) long from the current season's growth. Remove the tip, if it is still soft, cutting back to the ripened hardwood. Trim off any sideshoots. Make planting holes with a stake or metal rod, then drop the cuttings into the holes. Firm in.

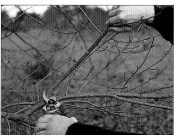

1 Select shoots (here of *Populus*) up to 2m (6ft) long from the current season's growth.

2 The cuttings are best rooted where they are to mature. Make planting holes by driving a rod into the ground to about 1m (3ft).

Hardwood Cuttings of Vines

Take hardwood cuttings of vines (*Vitis*) in late autumn. Short cuttings known as vine eyes containing one bud (or "eye") at the top can also be used. Make a cut immediately above a bud and another 5cm (2in) below it. Root all vine cuttings with bottom heat.

STANDARD CUTTING

VINE EYE

1 Standard hardwood cuttings of vines are taken with 3–4 buds. They root more readily than vine eyes.

2 When the cuttings break into bud in spring, pot them singly. Grow them on until the following spring before planting them out.

Basal Stem Cuttings

MANY PERENNIALS CAN PROVIDE material for basal stem cuttings from the first flush of new growth in spring. These cuttings consist of entire young shoots severed from the crown of the parent plant, so that each retains a piece of parent tissue at the base. If taken very early in the season from summer-flowering plants such as asters, salvias or phlox, these cuttings should make reasonably sized flowering plants by summer or autumn of the same year.

Basal Stem Cuttings from Stock Plants

Stock plants of some plants, such as chrysanthemums, can also yield basal stem cuttings, when they have been overwintered under cover. The stock plants are usually then discarded, because the new plants will have more vigour than the parent. A suitable propagating compost may be mixed from equal parts sand and peat or a peat substitute such as coir. Some additional bottom heat will improve chances of rooting when the cuttings are taken early in the year.

1 Cut new shoots cleanly through at the junction with the woody crown tissue, ensuring the base of each cutting has a small amount of this tissue attached. These chrysanthemum cuttings are 8–10cm (3–4in) long.

2 Remove the lower leaves and trim the bases, cutting straight across below a node if visible, or so the cuttings are 5cm (2in) long. Treat the base of each chrysanthemum cutting with hormone rooting powder or gel.

Always label pots of cuttings

Lower pair of leaves removed

3 Insert the cuttings into pots of a suitable cuttings compost. Water well and label clearly. Put the cuttings in a propagator or tent them in a clear plastic bag. As these cuttings are usually taken early in the season, bottom heat of 10°C (50°F) will speed rooting.

4 When well-rooted, usually after about four weeks, separate the cuttings. Aim to keep disturbance to the roots to a minimum. Pot the cuttings singly in standard potting compost (*see inset*). Stand them in a cold frame or greenhouse for growing on.

Basal Stem Cuttings in Perlite

Basal stem cuttings of many perennials can be taken from the first new growth in spring. Even earlier cuttings can be obtained by gentle forcing of plants that have been lifted and potted in the previous autumn (as with the delphinium shown here). Forcing involves starting them into early growth in a greenhouse or cold frame. Some plants, including diascias and violas, can be induced to form material suitable for basal stem cuttings later in the season: cut back flowered stems to the crown and top-dress it with gritty compost. Some perennials, notably lupins and delphiniums, have hollow stems that tend to rot in compost. Taking basal stem cuttings seals the stem against rot as the base of the cutting has a piece of crown attached. For hollow-stemmed cuttings, a light, open medium such as perlite or vermiculite helps to prevent rot; regularly spray or drench the cuttings with fungicide. Bottom heat of 15°C (59°F) improves rooting.

1 Select new shoots that are about 8–10cm (3–4in) long, in spring. Cut off at the base, keeping a piece of the parent's woody crown. Trim off all except the top 2–3 leaves. With a clean sharp knife, remove any stubs.

Stubs are removed from lower stem

2 Remove any damaged tissue. Fill a 15cm (6in) pot with moist perlite to within 2.5cm (1in) of the rim. Stand the pot in a saucer of water. Gently push in about 8 cuttings so that they are half-buried.

3 Label the pot and stand in its saucer of water in a warm place out of direct sunlight. Keep the perlite constantly moist. The cuttings should root in 4–8 weeks and are ready for potting when the new roots are about 1cm (½in) long. Ease them out gently.

4 Shake off any loose perlite. Pot the rooted cuttings singly into 8cm (3in) pots of soilless potting compost, at the same depth as before. Firm gently, label and water. Grow on the cuttings for 6–8 weeks until they are established before planting them out.

LEAF-BUD CUTTINGS

OFTEN TAKEN FROM SHRUBS AND CLIMBERS, leaf-bud cuttings make economical use of softwood and semi-ripe material from the parent plant, producing several cuttings from one vigorous shoot. A leaf-bud cutting requires only a short piece of stem to provide food reserves, since it also manufactures some food through its leaves. Take cuttings in spring or summer.

TAKING LEAF-BUD CUTTINGS

Leaf-bud cuttings can be internodal (cut between leaf joints), which usually works well with clematis and honeysuckle, or nodal (cut below a leaf joint), which is more suitable for plants with hollow stems or ones that are susceptible to rot, such as camellias. Cut off a strong shoot and sever it between the nodes to create a number of internodal cuttings, each with one or two leaves. Take care to retain the growth buds in the leaf joint at the tip; they are all too easily nipped out in error. With some species the buds are quite long; in this case the cutting should be cut back to a few millimetres above the top pair of leaves. You may need to wound very woody stems by cutting away bark from the stem base, but this is not generally necessary.

CAMELLIA CUTTING

LEAF-BUD CUTTINGS OF CLEMATIS

Leaf-bud cuttings of softwood are taken from spring to midsummer and of semi-ripe wood from mid- to late summer. They all root well but semi-ripe cuttings need less humidity. For large-leaved cuttings such as *Clematis montana* cultivars and *C. armandii*, reduce the cutting to a single leaf to avoid overcrowding and grey mould. Take internodal leaf-bud cuttings about 5cm (2in) long from the current season's growth. Look for well-formed buds; weak buds may not produce new shoots.

PREPARED CUTTING

STRONG BUDS

WEAK BUDS

ROOTING TIMES

- The time taken for a cutting to root depends upon the plant, the type of cutting, age of the stem, how it was prepared and the rooting environment.
- Leafy cuttings root in about three weeks; woody cuttings take up to five months. Leaf-bud cuttings generally take about eight weeks to root, depending on the woodiness of the stem.

PLANTS FOR LEAF-BUD CUTTINGS

Camellia japonica and *C. × williamsii* cultivars.
Clematis All species and cultivars.
Dracaena fragrans and cultivars, *D. marginata* 'Tricolor'. Variegated cultivars must be grown from cuttings to retain variegation.
Ficus elastica (rubber plant) and cultivars.
Hedera canariensis, *H. colchica* and *H. helix* cultivars (ivy).

Lonicera japonica, *L. periclymenum*, *L. sempervirens* and other climbing species (honeysuckle).
Mahonia aquifolium, *M. japonica*, *M. × media* and other species and cultivars.
Passiflora caerulea and other species (passion flower).
Philodendron scandens and other species.
Rubus cockburnianus and other species (brambles).

Leaf-bud Cuttings from Shrubs and Climbers

Leaf-bud cuttings are made up of a single leaf or a pair of leaves containing a growth bud and a short piece of stem. If appropriate for the plant (see *Taking Leaf-bud Cuttings* on facing page), divide a section of stem into nodal cuttings by cutting each one just below a leaf joint. If the plant has large leaves, such as *Lonicera* (honeysuckle), it is a good idea to cut them in half, to reduce moisture loss. Apply hormone rooting compound to the base of each cutting, shaking off the excess if using powder. Insert the cuttings around the edge of a pot filled with standard cuttings compost. After watering in with fungicide solution and labelling, keep the cuttings humid by placing them in a propagator or under a plastic film tent. Some less hardy subjects may need bottom heat to aid rooting. Cuttings take around eight weeks to root. Pot the young plants into single containers using potting compost and grow them on in protected conditions until established, then harden off and plant out.

1 Select a healthy shoot of the current season's growth (here of ivy, *Hedera*). Cut the stem just above every node to create inter-nodal cuttings with 1–2 leaves. Prepare nodal cuttings by trimming below a leaf at the base and above a leaf at the top.

2 Dip each prepared cutting (*see inset*) in some hormone rooting compound. Fill a pot with standard cuttings compost and make holes for the cuttings. Insert each cutting into the compost, so that the leaves are held just above the surface and do not touch.

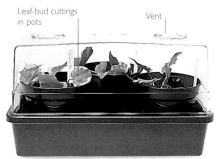

3 Firm and water in the cuttings and label the pots. Place the pots in a propagator and keep the environment humid by misting if needed. Bottom heat is generally not required for ivies. The cuttings should take about 8 weeks to root.

4 Pot the rooted cuttings individually in standard soilless potting compost, into pots about 1cm (½in) larger than the root ball of each cutting (*see inset*). Water in each cutting thoroughly and label. Grow on until the plants are well established.

Leaf Cuttings

Some plants can regenerate roots and shoots from part or whole leaves (although variegated plants cannot be used for this purpose). There are two types of leaf cutting: with the first, the whole leaf is used, and with the second, new plants form on the surface or cut edge of a sectioned leaf, as in *Streptocarpus* and *Sansevieria*. Leaf cuttings are usually taken early in the growing season.

Succulent Leaf Cuttings

Some types of succulent, for example many species of *Crassula*, *Kalanchoe* and *Echeveria*, may be propagated from leaf cuttings. Take the cuttings, selecting firm fleshy leaves, and pot them as shown. Place in a bright position, shielded from direct sun and keep slightly damp. The leaves should produce roots after two to four weeks. After one month, tiny new plantlets will develop.

1 Fill a pot with gritty compost topped with fine grit. Push the base of each leaf (here *Pachyphytum oviferum*) deep enough into the grit for the leaf to stand up.

2 Label and place in a bright, warm, airy position. Keep slightly moist. After 1–6 months, the leaves should have rooted and produced new plantlets.

Whole-leaf Cuttings

This method involves a whole leaf and its stalk, and sometimes a dormant bud at the base of the stalk where it joined the stem. On African violets (*Saintpaulia*) the bud is not crucial because a new one will form. The cuttings need a free-draining medium.

1 Cut healthy, mature leaves (here of African violet) from the parent plant, close to the base of the leaf stalk. Insert in pots of equal parts peat and coarse sand.

Improvised cloche

2 Water the cuttings, allow to drain. Cover to prevent moisture loss: here, clear plastic bottles are cut down to make cloches. Provide bottom heat and shade from sunlight.

3 Remove the covers after several plantlets have formed around each leaf base. Allow the new plants to grow on until they are large enough to be potted singly.

Part-leaf Cuttings

With this method, new plants form from the wounded veins of a sectioned leaf. *Streptocarpus* is shown here; sansevierias can be propagated from transverse sections (*see p.29*). Plantlets appear along the veins; when they are large enough, pot on.

1 Select a healthy, full-grown leaf and cut it into sections, so that the veins in the leaf are wounded. Here a *Streptocarpus* leaf is cut in half and the midrib discarded. Prepare a seed tray of free-draining cuttings compost.

2 Make shallow trenches in the compost and insert the leaf cuttings in them, cut side down. Firm gently around the base of the cuttings. Put the tray in a propagator or seal in a plastic bag to prevent moisture loss.

Leaf Cuttings from Begonias

Leaf cuttings root readily from *Begonia rex*, *B. masoniana* and similar types. Take leaf cuttings in spring or summer; plantlets should form in about six weeks to two months. Two different techniques are illustrated here.

1 Select a fully grown, healthy leaf (here of *Begonia rex*). Using a sharp knife, cut off the leaf stalk and then cut straight across each of the main veins on the underside of the leaf. Each cut is 1cm (½in) long.

2 Pin the leaf, cut side down, onto the surface of a tray of standard cuttings compost or vermiculite. Pin over the veins to keep them in contact with the compost. Label the tray. Keep humid at 21°C (70°F) until plantlets develop, in around two months.

3 Alternatively, cut squares, about 2.5cm (1in) across, from a large, healthy leaf. Each square must have a main vein running through it. Secure them with pins, veins downwards, in a tray of cuttings compost and treat as in Step 2.

Root and Rhizome Cuttings

A LIMITED RANGE OF PLANTS – ones that naturally produce shoots, or suckers from the roots, such as *Rhus typhina*, can be propagated from root cuttings. The roots are usually thick and fleshy, as they store food for producing shoots. Some plants, such as pulsatillas, grow well from root cuttings but there will be a check in growth from the root disturbance. Cuttings can also be taken from mature rhizomes (swollen underground stems).

Taking Root Cuttings of Acanthus

Acanthus species increase naturally from roots left in the soil, and are easy to propagate from root cuttings. Lengths of healthy root can be taken in the dormant season, of pencil or medium thickness for the plant. Cuttings flower in two years.

1 Lift the plant in autumn or winter when it is dormant and wash the roots. Choose strong roots, of medium thickness for the plant, and sever them, cutting as close to the crown as possible. Remove no more than one-third of the available root material.

2 Cut each root into sections that are 5–10cm (2–4in) long; making the thinner cuttings the longest. To make sure that you insert the cuttings the right way up, cut the base of each cutting at an angle and cut the top of each cutting straight across (*see inset*).

3 Prepare pots of moist cuttings compost. Treat the cuttings with a fungicide. Use a dibber to make holes as deep as the cuttings and insert them vertically. The top of each should be level with the compost surface. Top-dress with a layer of coarse sand or grit.

4 When new top growth appears, usually by the following spring, gently tease out the cuttings and check for root growth. When ready, pot the cuttings individually in 8cm (3in) pots filled with standard potting compost. Water well and label the pots.

Cuttings of Culinary Herbs

Herbs with creeping roots such as horseradish, or rhizomes such as mint, can be propagated by root or rhizome cuttings. Take the cuttings in spring or autumn. Lift the parent plant and remove some healthy roots or rhizomes. With most herbs, including mint, they should be of average thickness. Horseradish roots root readily whichever way up they are, so can be sliced into small sections. Keep the cuttings in a bright place, but out of direct sunlight. Do not water until new roots or top-growth appear (2–3 weeks).

Roots are sliced into 1cm (½in) sections

1 In spring, lift a healthy horseradish plant, taking care not to damage the roots. Cut off 1–2 lengths of root, 15–30cm (6–12in) long. Prepare a container with cuttings compost and firm.

2 Slice the roots into 1cm (½in) sections. Insert each cutting 2.5–6cm (1–2½in) deep in the prepared module tray. When the cuttings have good root systems, transplant to their final positions.

Taking Rhizome Cuttings of Bergenia

Rhizomes are usually swollen underground stems, either thick, as in bearded irises, thin and fast-growing, as with wild rye (*Elymus*), or in a crown, as in asparagus. As a rhizome grows, it often develops segments, each with buds that break into growth when conditions are favourable. These can be cut apart for propagation.

Bergenias, mostly fully hardy perennials, form woody, creeping rhizomes, often on the soil surface, with leaves only at their tips. Detach these, with some roots, if just a few plants are required. To produce large numbers of new plants, take rhizome cuttings. Cut the older part of the rhizomes, which are devoid of leaves, into sections.

1 Cut older pieces of leafless rhizome into 4–5cm (1½–2 in) sections, each with several dormant buds, in autumn. Trim any long roots. Half-bury the sections, buds uppermost, about 5cm (2in) apart in trays in moist perlite or compost. Water and label.

2 Keep the cuttings at a humid 21°C (70°F) in a heated propagator, shaded to prevent dehydration. After 10–12 weeks, plantlets (here of *Bergenia cordifolia*) should have rooted. Pot singly or line out in a nursery bed, in spring. Flowering occurs in 1–2 years.

More Specialist Cuttings

THESE VARIATIONS ARE USED to propagate more specialist plants. Rosette cuttings are taken from some, including cushion-forming alpines and various succulents. Certain shrubs, trees and perennials, including *Yucca*, *Cordyline* (cabbage palm) and *Dieffenbachia* (dumb cane) are increased from leafless stem sections. Stem cuttings without leaves can be used to propagate some columnar and epiphytic or forest cacti.

Rosette Cuttings in Ground Pumice

Cuttings can be taken from rosette-forming alpines in late spring and in summer. Handle parent plants with care, for they are easily damaged. The short-stemmed cuttings also need careful handling. Use ground pumice if possible for difficult alpines. Derived from Icelandic volcanic rock, it is totally sterile and sufficiently water-retentive for alpines. Rooting of rosette cuttings is slow and spasmodic, however.

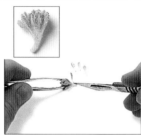

1 Select a healthy rosette from the edge of the plant (here *Dionysia aretioides*). Cut the stem 5–10mm (¼–½in) below the shoot tip. Trim lower leaves and dip base into hormone rooting compound.

2 Fill a 5cm (2in) clay pot with ground pumice to within 1cm (½in) of the rim. Water from below and allow to drain. Insert the cuttings, 1cm (½in) apart. Firm around them and label the pot.

Leafless Stem Sections

Various plants, such as stem-producing yuccas, cordylines and dieffenbachias, can be increased from leafless stem sections. Cuttings of all are usually inserted vertically into the rooting medium, but yucca cuttings can be laid flat to induce young shoots for use as softwood cuttings. The cut-back stems of parent plants will re-shoot.

1 Remove a 30–90cm (12–36in) section from a mature stem (here *Yucca elephantipes*), cutting between the leaf nodes. Strip all the foliage. Cut the stem into 10cm (4in) cuttings.

2 Trim alternately above and below a node. Press the cuttings flat into moist soilless cuttings compost, so they are half buried, or insert vertically into 9cm (3½in) pots.

Succulent Rosette Cuttings

Some rosette-forming succulents consist of clumps of rosettes. These can be severed at the base, and rooted as shown here. Once potted, keep the cuttings in a bright, airy position.

1 Using a sharp clean knife, cut 5–8cm (2–3in) from the top of a young rosette of leaves (here of *Echeveria* 'Frosty'). Trim off the bottom leaves (*see inset*) and allow to callus for a few days.

2 Gently push the stem through the fine grit top-dressing into the compost, so the leaves sit just above the surface. Use a standard 8cm (3in) pot or a deep seed tray.

Flat Stem Cuttings from Cacti

Some epiphytic, or forest, cacti such as epiphyllums and Christmas cacti (*Schlumbergera*), will root from sections of their flat, leaf-like stems. These cacti generally prefer a more humid environment than desert types, and prefer partial shade. Take a whole mature stem from the parent plant, at the base, in late spring or early summer, after flowering.

Tip of stem is top of cutting

1 Cut a flattened stem (here of an *Epiphyllum*) into 23cm (9in) sections. Allow to callus for a few days. Fill a pot with cactus compost.

2 Fill the pot to just below the rim. Cover the compost with a shallow layer of fine grit and push the cutting into the compost below.

Stem Cuttings of Columnar Cactus

Cuttings up to 2m (6ft) long can be taken from columnar cacti such as the monstrose form of *Cereus hildmannianus*, shown here. The larger the wound on the cutting, the longer it will take to callus. Cuttings should root in 1–12 months.

1 Wear thick gloves and use a folded cloth to steady the plant. Use a large knife to remove a 8cm–1m (3in–3ft) length.

2 Leave the cutting in a warm, dry place to callus (2–3 weeks in summer). Fill the bottom third of a pot with cactus compost and add a layer of fine gravel. Stand the cutting on this and fill with gravel to the top. Keep slightly moist.

Caring for Cuttings

Propagating under Cover

ONCE ANY PLANT MATERIAL has been correctly prepared for propagation, and inserted into a suitable growing medium, it is important to provide conditions that will enable the material to survive and establish. The degree of care needed depends on the species of plant and the mode of propagation. Hardwood cuttings, for example, require minimal care, but leafy cuttings taken in summer from a difficult plant will need a closely regulated environment.

Providing the Right Support

Propagation involving regenerative processes such as the formation of new roots or shoots immediately demands some form of environmental support until the new plants become independent. In cooler climates, favourable conditions can often only be achieved under cover – for instance, in the home, conservatory or greenhouse – to extend the growing season or increase tender plants. For outdoor propagation, cold frames, cloches or nursery beds offer a degree of shelter. In warmer regions, windbreaks, shading structures and irrigation systems may be required.

Propagating plants away from their natural or adapted habitat makes them vulnerable to attacks from pests and diseases, so the propagation area should be kept as clean as possible.

In general, cuttings need water, warmth, air, light and sometimes nutrients to grow. The humidity of the air is a critical issue for many unrooted leafy cuttings, for example. In spring and summer they may need 98–100 per cent humidity to prevent wilting. Plants are temperature-dependent and grow best when warm. Ventilation helps to prevent fungal infection such as botrytis, but must also be regulated to avoid excessive loss of humidity.

"TENTING"
The easiest way to keep humid air around cuttings is to create a "tent" with a clear plastic bag. Hold the bag clear of the plant material with a wire hoop or a few split canes.

◀ CONTROLLED ENVIRONMENT *A necessity for many cuttings, easily achieved in a greenhouse.*

WINDOWSILL PROPAGATION

The simplest propagation environment can be created by keeping individual containers on a bright windowsill or in a glassed-in porch or conservatory. The location provides warmth and light; humidity is maintained by covering the container. For a pot of cuttings, use a plastic bag or a bottle cloche. Whatever type of cover is used, it must be kept clear of the plant cuttings.

▲ RETAINING A HUMID ATMOSPHERE
Propagated material such as leafy cuttings must often be kept in a contained space to keep the air humid. Moisture from the cuttings rises to the top of the propagator. The cover stops that moisture evaporating and the vent allows excess humidity to be controlled.

Lid redirects moist air back to the plants and maintains humidity

◄ HOME GROWN
It need not be difficult to root cuttings at home. A bright windowsill is suitable for cuttings.

► HEATED HOME
Portable propagators can be used indoors. Some are fitted with electric heating elements to provide bottom heat.

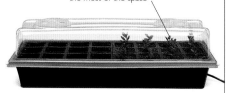

Modular inserts make the most of the space

CHOOSING A GREENHOUSE

A greenhouse is a valuable asset in cool climates, allowing varying degrees of environmental regulation. There are many different styles available. Some models are designed for maximum light penetration, heat conservation or ventilation, while others make the best use of restricted space.

There are four categories: cold (unheated), cool, temperate, and warm; the choice depends on the plants you want to grow.

A lean-to or mini-greenhouse benefits from the warmth and insulation of the house wall, but may suffer more from extreme temperature changes.

PLASTIC TUNNEL
This low-cost greenhouse is covered with heavy-duty film plastic, which will eventually become opaque.

MINI-GREENHOUSE
This is very useful where space is limited. Choose a position carefully for maximum heat and light.

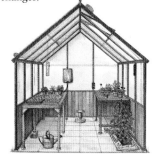

TRADITIONAL GREENHOUSE
This allows the most complete and controlled propagation environment. Different areas can be created for various uses.

THE PROPAGATOR'S GREENHOUSE

A greenhouse provides the gardener with the chance to create various separate environments for propagating and growing plants. This greenhouse is equipped with all the elements necessary to propagate a wide range of plants. Insulation or heating may be needed in winter.

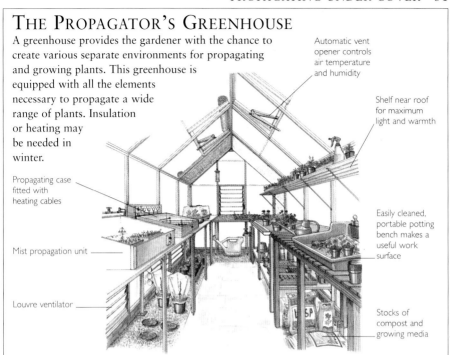

PROVIDING AN EXTRA BOOST FOR CUTTINGS

The temperature of the growing environment will affect plant growth. A heated propagator provides thermostatically controlled bottom heat, to ensure that the growing medium is warmer than the air above. Soil-warming cables, sold in varying lengths and wattages, can be used. More sophisticated systems such as mist and fog propagators (*see p.13*) offer the additional bonus of regulated air humidity.

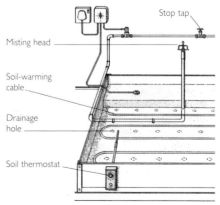

USING SOIL-WARMING CABLES
Lay the cable in a series of "S" bends in a bed of moist sand at a depth of 5–8cm (2–3in), making sure the loops do not touch.

INSIDE A MIST PROPAGATOR
As well as supplying bottom heat through an electrically heated bed of sand or compost, a mist of fine water droplets is sprayed.

Cold Frames and Cloches

In the open garden in cooler climates, cold frames and cloches may be used to warm the soil and air, increase local humidity, give shelter from drying winds and some protection from pests. They can provide a suitable rooting environment for a wide range of easily rooted cuttings. Cold frames are more permanent structures than cloches. They may also be used for overwintering rooted cuttings and for the hardening off of new plants.

Different Types of Cloche

Cloches provide a suitable rooting environment for a variety of easily rooted cuttings. A wide range of designs and materials is available. The best are glass or plastic; plastic allows less light penetration and retains less heat. Plastic film and rigid polypropylene last five years or more; rigid, twin-walled polycarbonate lasts for at least ten years. Well-fitting end pieces are essential to stop the cloche becoming a wind tunnel. Rigid cloches are more costly but more easily moved, making watering and transplanting easier. Some are self-watering, with permeable coverings that allow rainwater to trickle through.

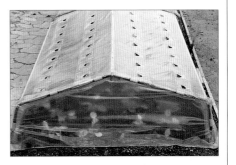

▶ SELF-WATERING CLOCHE
This type of cloche is designed to allow rainwater, or water from a garden sprinkler, to trickle through to the plants or cuttings below.

PLASTIC FILM TUNNEL
Sturdy wire hoops are covered by plastic film, which allows easy accessibility. Tying the plastic prevents the interior from becoming a wind tunnel.

RIGID PLASTIC TUNNEL
This type of plastic cloche can be any length and is held in position by a metal or plastic frame. The rigid plastic should last 5 years or more.

TENT CLOCHE
Simply constructed from two sheets of glass to form a tent shape, this kind of cloche can be used singly or in a row to cover low plants and cuttings.

The Advantages of a Cold Frame

Cold frames provide propagation material and new plants with higher soil and air temperatures, reduced temperature fluctuation if shaded and ventilated, shelter from winds and adequate light levels. Cold frames also suit certain plant material such as grey-foliaged Mediterranean plants or hardwood cuttings, which do not like the humidity of a closed case or propagator. Cuttings can be inserted directly to root in a nursery bed in the frame. A good number of pots or trays can be accommodated in a cold frame, and soil-warming cables (see p.51) may also be used in the bed.

◀ MOVEABLE COLD FRAME
This portable cold frame may be moved around to take advantage of seasonal light.

▲ FIXED COLD FRAME
This brick frame (with glass removed for sunny days) holds heat and excludes draughts.

Using a Cold Frame

Cold frames with metal frameworks let in most light and can be moved around the garden to follow the best light at different times of year, but they do not retain heat or exclude draughts as well as timber and brick frames. Permanent frames must be sited in a sheltered position, where maximum light is received all year round.

Cold frames overheat in sun unless they are ventilated and shaded. If the temperature falls below –5°C (23°F), insulate the frame to avoid frost damage.

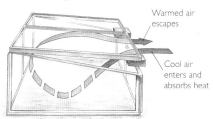

AIR CIRCULATION
Cold air expands and rises as it heats up on a warm day. Open the lights of the frame to allow some warm air to escape; the inside temperature should remain reasonably cool.

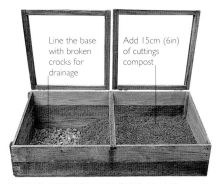

FILLING A PERMANENT COLD FRAME
A fixed frame can provide a nursery bed for cuttings. Cover the base with a thick layer of drainage material, such as broken crocks or coarse gravel. Top up with cuttings compost.

Care of Cuttings under Cover

THE CHOICE OF GROWING MEDIUM should provide the propagated material with an appropriate amount of oxygen and nutrients, but correct watering and temperature control of the medium is also needed for the various growth processes to occur (such as root initiation). Long summer days help with propagation, but intense light will overheat the air, in turn causing excessive transpiration and stress to cuttings. Shading to create indirect light aids rooting.

Importance of Warmth and Humidity

The humidity of the air affects the rate at which plants transpire, or allow water to evaporate from leaf pores. The more humid the air, the less the plants transpire (see p.50). This is a critical issue for unrooted leafy cuttings, which in spring and summer need a very humid atmosphere to prevent wilting. Wilted cuttings have a reduced ability to regenerate, form callus tissue or develop roots. Cuttings absorb moisture through their severed bases more quickly than through leaves but once callus tissue forms, water can only be taken in by the leaves. Excessive transpiration can cause stress in cuttings, resulting in leaf-drop, but can be controlled by shading (see below).

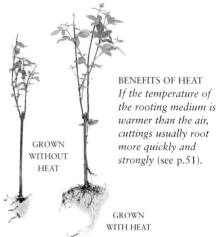

GROWN WITHOUT HEAT

GROWN WITH HEAT

BENEFITS OF HEAT
If the temperature of the rooting medium is warmer than the air, cuttings usually root more quickly and strongly (see p.51).

Ventilation and Shading

Shading should protect plant material from being scorched by direct sun while still allowing sufficient light to pass through it for good growth. Direct sunlight heats the air, causing stress in cuttings due to excessive transpiration. Some shading materials used on greenhouses include shading washes, while shading netting can be used on smaller structures. Excess heat can also be controlled by ventilating the propagating environment.

AUTOMATIC VENTING
Adequate ventilation is essential to control air temperature and humidity in a greenhouse. An automatic vent opener controls both these elements by opening automatically when the temperature rises above a predetermined level.

▲ SHADING WASH
In warm weather, shading will help control air temperature in a greenhouse. Washes make very effective shading because they reduce the heat from the sun significantly while allowing through enough light for good growth. Apply the wash externally.

WATERING AND AFTERCARE

It is important to provide a regular and adequate supply of water for cuttings, especially in hot or windy weather, and particular attention should be paid to plants growing in smaller containers, which can dry out quickly. Adequate water is crucial for good plant growth; however, an excess of water (waterlogging) can be just as damaging as lack of water, as it will deprive the roots of oxygen and promote rot. Possibly more cuttings are lost in this way than for any other reason.

FINE BRASS ROSE

WATERING IN CUTTINGS
Use a fine rose turned upwards to water cuttings (here of rosemary). It creates a fine, light spray and avoids disturbing the compost.

AVOIDING PLANT PROBLEMS

Propagated plants live in an artificial environment that leaves them vulnerable to attack from harmful pests and diseases. The use of bottom heat, frequent watering and high humidity that are so often essential in propagation also encourage the proliferation of debilitating fungal diseases, for example. These are often introduced through poor hygiene in the preparation of plant material or through contaminated composts. Pests and diseases can also be transmitted from the parent plant; this can be a particular problem with pests that are not easily discernible, such as eelworm.

Regularly check new plants and control any problems as soon as they arise. Cuttings that show signs of rot, viruses or frost damage should be discarded as soon as the problem is noted. It is best to try and prevent plant problems occurring in the first place. Observe good hygiene, use sharp, clean instruments for cutting, and use sterile growing media. Greenhouses should be scrubbed annually with a solution of horticultural disinfectant; this will help to control sciarid and whiteflies, red spider mite, mildew and the various fungi that cause damping off or blackleg.

BLACKLEG
Before or as roots form, the base of a cutting darkens and atrophies; the upper parts discolour and die. It is caused by soil or water-borne fungi.

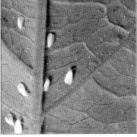
GREENHOUSE WHITEFLY
These sap-feeding insects cause stunted growth and distorted leaves. Use insecticidal soaps and/or biological control.

GREY MOULD
Grey mould (Botrytis cinerea) thrives in damp conditions and spores may persist from year to year. Increase air circulation around plants.

The Rooted Cuttings

Once the cuttings have rooted, they need special care to get them acclimatized to the growing environment. They must be slowly weaned from the cossetted conditions in which they have been raised; they will need potting, and plants intended for outdoors must be hardened off in a cold frame.

Weaning Propagated Plants

Once plants have functioning root and shoot systems that are adequate for independent survival, the process of weaning them from the propagation environment into a growing environment should take place. Cuttings that have been rooted in closed propagating cases, under plastic film on a heated bench, or in mist or fog propagation units need particular care. It takes two to three weeks for plants to fully acclimatize. First any bottom heat is turned off. The humidity level is then gradually reduced; plastic film and propagator covers are removed for a longer period each day until eventually they are not replaced at night; and the duration and frequency of mist or fog bursts are reduced, then the units are switched off at night. Once weaned, place the new plants in temperatures appropriate to the species.

Growing on heaths and heathers, as shown here, is a good example of how to treat many other hardy shrubs once the cuttings have rooted under cover.

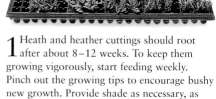

1 Heath and heather cuttings should root after about 8–12 weeks. To keep them growing vigorously, start feeding weekly. Pinch out the growing tips to encourage bushy new growth. Provide shade as necessary, as direct sun causes stress in young plants and scorches foliage. Ventilate plants daily.

Bushy young plants ready for transplanting

2 The well-rooted cuttings are potted individually into 8–9cm (3–3½in) pots, using ericaceous compost for lime-hating heathers. Grow on outdoors after hardening off in a cold frame. The hardening-off process should take place over a period of two to three weeks.

3 When large enough, plant out in flowering positions. A good time for planting heathers and other young shrubs is autumn while the soil is still warm. This encourages plants to rapidly make new roots and become established before cold weather sets in. The alternative is spring planting.

POTTING TENDER PLANTS

When cuttings are well rooted, carefully transfer each one singly into a pot filled with potting compost. The pot should not be too large. Colourful-leaved coleus (*Solenostemon*), shown here, are popular and not difficult from cuttings.

Four-week old cuttings

1. The cuttings have been placed in rockwool modules (*see inset*) in a propagator to keep humid, in bright light at a minimum temperature of 18–21°C (64–70°F). Roots develop after about two weeks.

2. Pot the rooted cuttings singly into 9cm (3½in) pots of soilless potting compost. Do not tease out the roots from rockwool modules. Grow on the young plants in a warm, bright place.

HARDENING OFF AND PINCHING OUT

Before planting out, young plants raised under cover must be hardened off by acclimatizing them to the temperatures outdoors. This may take two to three weeks and must not be rushed, because the plant's leaves must undergo changes to reduce water loss. Transferring the young plants to a cold frame is ideal – ventilate gradually until the covers are fully open at night as well as by day. A cloche is an alternative, but does not offer as much frost protection. Encourage bushy growth by pinching out growing tips.

▶ PINCHING OUT
Pinching out the growing tip encourages sideshoots and a bushier plant.

OPENING HINGED LIGHTS (COVERS)
A cold frame is ideal for hardening off. Hinged lights can be propped open to stop plants overheating, but may admit winds.

OPENING SLIDING LIGHTS
Sliding lights can be removed entirely if desired, but this would leave plants unprotected in the event of heavy rain.

Cuttings in the Open

Some cuttings, particularly many hardwood cuttings of woody plants, are rooted outdoors direct in the soil. Very often, after hardening off, young hardy plants are planted out in a nursery bed to grow on for a year or two. For both of these you will need to thoroughly prepare some beds in a sheltered position, possibly provide protection from the elements, prevent them from drying out and watch out for a number of pests and diseases.

Simple Digging

Soil beds intended for hardwood cuttings or young plants are prepared by single digging (to one depth of the spade blade), which is best done several months before the beds are needed. During the digging, add plenty of bulky organic matter, such as garden compost, to each trench to help the soil retain moisture. If the drainage is poor, also add grit to each trench and mix some into the topsoil.

1 Simple digging just involves turning over spadefuls of soil. Drive the spade down to its full depth, keeping the blade upright. Press down firmly with the ball of your foot.

2 Pull back on the handle and lever soil on to the blade. Bend your knees and elbows to lift the spade; do not try to lift too much at once, especially where the soil is heavy.

Protecting Young Plants

Outdoor nursery beds do not have the controlled environment found under cover but may need some form of protection. Drying winds can stress plant material by increasing moisture loss: construct windbreaks on the side of the prevailing winds or use cloches. In warm climates or dry seasons, beds may also need irrigation: seep hoses are useful and can be laid alongside new plants.

WIND PROTECTION
To protect cuttings and young plants from cold, drying winds, erect a screen of windbreak netting on the windward side. This filters and slows down the wind.

SHADING
Supporting plastic shading netting over young plants prevents new growth from being scorched by the sun, particularly during hot summers or in warm climates.

CUTTINGS IN THE OPEN • 59

DEALING WITH PROBLEMS

Most plant problem symptoms are easy to see, from mould to visible insects. Check plants frequently to spot any problems as soon as they arise. Do not allow young plants or cuttings to dry out and make sure the soil has adequate nutrients for healthy growth. Observe good hygiene and try to use organic methods of control before resorting to chemicals.

CUTWORMS
These gnaw on roots and stem bases and feed on the soil surface at night. Find all of the cutworms and destroy.

POWDERY MILDEW
White, powdery fungal growth appears on leaves. Remove infected leaves immediately and avoid overhead watering.

FROST DAMAGE
The upper parts of leaves turn brown or black. Nip off affected leaves or discard badly damaged plants.

SLUGS
These feed on soft plant material, mainly at night or after rain. Hand-pick or use slug pellets, barriers or traps.

VINE WEEVIL LARVAE
These grubs feed on roots. Plants grow slowly, wilt and die. Use biological control or insecticidal compost.

IRREGULAR WATERING
Adequate water is crucial for good plant growth, and cuttings must be regularly watered to avoid wilting.

APHIDS
These sap-feeding insects cause stunted growth and distorted leaves. Control with insecticidal soap.

IRON AND MANGANESE DEFICIENCY
Leaf discoloration means plants may need acid soil. Water with sequestered iron.

VIRUSES
Many viruses are transmitted from infected plants by sap-feeding insects. Destroy all affected plants promptly.

GARDEN TREES • 61

Plants to Grow from Cuttings

A wide range of plants can be grown from cuttings – trees, shrubs, climbers, perennials, including alpines, and cacti and succulents. You can collect cuttings from your own plants, or perhaps exchange them with gardening friends.

☼ *Prefers full sun* ◐ *Prefers partial shade* ● *Tolerates full shade* ◊ *Prefers well-drained soil* ◑ *Prefers moist soil* ◆ *Prefers wet soil* ✿ *Frost tender (min. 5°C/41°F)* ❋ *Half hardy (min. 0°C/32°F)* ❋❋ *Frost hardy (min. -5°C/23°F)* ❋❋❋ *Fully hardy (min. -15°C/5°F)* ♥ *RHS Award of Garden Merit*

Garden Trees

SMALL ORNAMENTAL TREES help to form the permanent structure of a garden. They are expensive to buy but many are surprisingly easy to propagate from cuttings, so you could have a small group of young trees or even grow some as shrubs by coppicing them regularly.

Acer
(Maple)
Moderately easy to propagate from softwood cuttings in early summer. Take cuttings from fast-growing tips of new shoots and make sure they do not wilt. Remove growing tip if cutting is longer than 10cm (4in). Root in a closed propagating case with bottom heat of 18–24°C (64–75°F).
☼–◐◊ ❋❋–❋❋❋

Ailanthus
(Tree of heaven)
Easy to propagate from root cuttings in early winter. Choose roots about 1cm (½in) in diameter but do not wash them. Root in a cool place, such as a slightly heated greenhouse. When rooted, in 3–4 months, line out in nursery bed or pot on. This tree can be coppiced.
☼–◐◊ ❋❋❋

Betula
(Birch)
Moderately easy from softwood cuttings taken from mid-spring to early summer. Root in a closed case with bottom heat of 18–24°C (64–75°F). Feed regularly once they have rooted to encourage maximum growth in first season, otherwise they may fail to start into growth in the following spring.
☼–◐◑ ❋❋❋

Acer palmatum 'Ôsakazuki' ♥

◀ A CLASSIC SHRUB BORDER *A scene like this could be created from cuttings*

CORDYLINE AUSTRALIS 'VARIEGATA'

Cordyline
(Cabbage palm)
Easily propagated from leafless stem sections at any time. Prepare from strong side shoots, each section with one or two nodes. Insert in shallow pans or half pots. Roots in 8–12 weeks in a warm propagating case, 18–21°C (64–70°F). It takes 3–5 years to obtain a good-sized plant.
◧–◧◊✿ (min. 13°C/55°F)–❋

Cornus
(Dogwood)
Moderately easy to propagate from softwood cuttings taken in late spring or early summer, as for *Acer* (see p.61). This technique is recommended for variegated cultivars. Pot up cuttings as soon as they have rooted. New plants take at least five years to flower.
◧–◧◊ ❋❋–❋❋❋

Ficus
(Fig)
Easily propagated. Take 30-cm (12-in) hardwood cuttings of *F. carica* in late autumn or winter. Root in pots at 10–15°C (50–59°F). Semi-ripe cuttings of tender evergreens may be taken all year, and rooted in a propagating case at 18–21°C (64–70°F). Species with thick stems (eg *F. elastica*) can be grown from leaf-bud cuttings – roll the leaf into a cylinder to reduce moisture loss. Keep humid at 20°C (68°F).
◧–◧◊✿ (min. 15°C/59°F)–❋❋❋

Ilex
(Holly)
Moderately easy. Take hardwood cuttings, 8cm (3in) long, with a basal wound, from autumn to midwinter. Provide bottom heat. Take semi-ripe cuttings, with a basal wound, in late summer to autumn. Both can take three months to root. Take semi-ripe cuttings of deciduous species in early or midsummer – do not wound. Root semi-ripe cuttings in a cold frame.
◧–◧◊ ❋❋–❋❋❋

Laburnum
(Golden rain)
Easy from 20–30-cm (8–12-in) long hardwood cuttings in late autumn. Remove with a heel or at the union of the current and last season's growth. Cutting into pithy tissue of new growth hinders rooting. Root outdoors, lining a trench with coarse grit, or in bundles in a cold frame, then pot in spring.
◧◊ ❋❋❋

Magnolia
Moderately easy. For deciduous species, take softwood cuttings from late spring to early summer, or greenwood cuttings from early to midsummer. Trim all but top two leaves. Root in humid shade with bottom heat of 18–21°C (64–70°F). Liquid feed rooted cuttings to ripen them and overwinter in a frost-free place.
◧–◧◊–◊ ❋❋❋

Morus
(Mulberry)
Moderately easy from 20-cm (8-in) hardwood cuttings in late autumn. Root them in a sheltered spot outdoors and lift a year later. Alternatively, take thick pieces of two- to four-year-old wood (truncheons) and root them outdoors – in final positions if desired. This will give larger young plants in a shorter period of time.
◧◊ ❋❋❋

Prunus
(Including cherries and plums)
Hardwood cuttings (30cm/ 12in) of deciduous species in late autumn are easy but slow. Insert in a sand bed in a cold frame over winter, in bundles of 10. Plant out singly in a nursery bed in spring (they should have rooted). Semi-

LABURNUM × WATERERI 'VOSSII' ♛

Magnolia denudata ♀

ripe cuttings of evergreens such as *P. lusitanica* ♀ from early to mid-autumn are moderately easy in a temperature of 20°C (68°F).

Robinia
(Locust)
Moderately easy from root cuttings taken from late autumn to early winter. Choose roots about 1cm (½in) in diameter but do not wash them. Cuttings can be 8–15cm (3–6in) in length. Store them vertically in a box of moist sand, in a cool, frost-free place. In early spring insert them 1cm (½in) deep in pots of free-draining compost to root at 10°C (50°F). Grow on the young plants in pots.

Salix
(Willow)
Very easy from hardwood cuttings (20cm/8in) taken from late autumn to early spring. Cuttings of vigorous willows may be as long as 2m (6ft) and planted out immediately to mature faster. Line out in a nursery bed.

CONIFERS

Mainly evergreen, conifers provide interest all year round and are excellent companions for deciduous trees. Some are also widely used for hedging.

Chamaecyparis
(Cypress)
Easy from semi-ripe cuttings taken from late summer to mid-autumn. Make them 10–15cm (4–6in) long: the base should not be too woody. Keep humid with bottom heat of about 20°C (68°F) maximum. Rooting may take 6–9 months.

Cupressus
(Cypress)
Cultivars are moderately easy to propagate from semi-ripe cuttings. For best results take 8–10cm (3–4in) green shoots in late winter and root them under mist with bottom heat of 20°C (68°F) maximum. Cuttings may also be rooted under cover in late summer. *C. macrocarpa* and cultivars are often used for hedging.

Juniperus
(Juniper)
Moderately easy from semi-ripe cuttings (10cm/4in). In late summer or autumn take strong, juvenile shoots (with needle-like leaves), green at the base. Root in a cold frame (should root within a year). Or, root in late winter with bottom heat.

Picea
(Spruce)
Challenging. From young plants, choose nearly ripe shoots, with firm but not woody bases. For mid-winter cuttings provide bottom heat of 15–20°C (59–68°F).

Podocarpus
Easily grown from semi-ripe cuttings in late summer. Can be rooted in a cold frame at that point; or root under mist or plastic film with bottom heat of 20°C (68°F).

Taxus
(Yew)
Fairly easy from autumn cuttings, 10–15cm (4–6in) long, from one- to three-year-old, upright, nearly ripe shoots, green at the base. Rooting by early summer outdoors; earlier with mist and bottom heat of 20°C (68°F).

Podocarpus nivalis

Shrubs and Climbers

Many of the plants used in creating the permanent framework of planting schemes are shrubs, while climbers can provide colour and interest at a higher level. When several shrubs are needed to form groups, the majority can be propagated from cuttings – and the same is true of climbers.

Argyranthemum
Easily raised shrubs. Usually propagated annually for summer bedding, from semi-ripe cuttings taken from non-flowering shoots in late summer. Root with gentle bottom heat, 15°C (59°F), in airy conditions. Then transfer to 9cm (3½in) pots of potting compost and winter in a cool, frost-free greenhouse.
◎ ◊ ❀ – ❀❀

Aucuba
Easily propagated shrubs. In late summer root semi-ripe cuttings in a sheltered nursery bed, such as a cold frame or under cloches. Alternatively, bottom heat at 21°C (70°F) speeds up rooting, which will take around 6–8 weeks. Leave cuttings until spring before potting. Plants in 3–4 years.
◎ – ◐ – ■ ◊ ❀❀❀

Argyranthemum 'Jamaica Primrose' ♀

Berberis
(Barberry)
Moderately easy shrubs. Take semi-ripe cuttings from mid-summer, or mallet cuttings in early summer (deciduous species) or autumn (evergreens). Take evergreen hardwood cuttings from late autumn to midwinter. Root in cold frames; use bottom heat for deciduous mallet cuttings.
◎ ◊ ❀❀ – ❀❀❀

Bougainvillea
Climbers, moderately easy to propagate. Softwood or semi-ripe cuttings (5–8cm/2–3in) in summer, taken with a heel, will root in 4–6 weeks if kept humid with bottom heat of 15°C (59°F). In winter root hardwood cuttings (20cm/8in) in deep pots with bottom heat of 21°C (70°F).
◎ ◊ ❀ (min. 5°C/41°F) – ❀❀

Buddleja
(Butterfly bush)
Easily propagated shrubs. Hardwood cuttings (20cm/8in) are simplest, taken from autumn to midwinter. Insert in deep pots and place in a cool greenhouse over winter. Plant out the following spring.
◎ ◊ ❀❀ – ❀❀❀

Callistemon
(Bottlebrush)
Easily propagated shrubs needing acid to neutral soil. Root greenwood to semi-ripe cuttings, prepared from side shoots, in summer and autumn. Take with a heel to aid rooting. Cuttings root well in rockwool modules. Provide steady bottom heat; 21°C (70°F) is ideal.
◎ ◊ ❀ (min. 5°C/41°F) – ❀❀❀

Camellia
Acid-loving shrubs, moderately easy to propagate from semi-ripe cuttings between midsummer and early autumn. They may be internodal or nodal leaf-bud cuttings with a 1.5cm (⅝in) basal wound, or nodal tip cuttings. Root in humid conditions with bottom heat of 20°C (68°F).
■ ◊ ❀ – ❀❀❀

Campsis
(Trumpet vine)
Climbers. Easy to raise from hardwood cuttings (20cm/8in) taken between autumn and mid-winter. Keep them cool and humid, eg in a cool greenhouse. Also easy from root cuttings (5cm/2in) in winter – they will root in a cold frame but respond more quickly in a frost-free greenhouse.
◎ ◊ ❀❀ – ❀❀❀

Caryopteris
(Bluebeard)
Easily propagated shrubs. Softwood cuttings from spring to midsummer and

SHRUBS AND CLIMBERS • 65

CHOISYA TERNATA ♀

greenwood cuttings from late spring to midsummer. In a warm humid environment under glass, rooting occurs within three weeks. Semi-ripe cuttings from mid- to late summer, as above or in a cold frame. Hardwood cuttings (20cm/8in) from late autumn to midwinter, in pots with gentle bottom heat.
◐◊ ✤✤–✲✲✲

Ceanothus
(California lilac)
Shrubs, moderately easy to grow from cuttings. Grow deciduous shrubs from softwood cuttings, taken in late spring to midsummer, with bottom heat of 15°C (59°F). Grow evergreens from semi-ripe cuttings with a heel, taken from midsummer to late autumn, as above. Pot root cuttings (5cm/2in) in autumn; place in frost-free greenhouse.
◐◊ ✤✤–✲✲✲

Chaenomeles
These shrubs are easily raised from hardwood cuttings, 20cm (8in) long, taken from autumn to midwinter. Wound the base of each cutting and root in pots in a cool greenhouse. Root cuttings taken from autumn to midwinter are also easy, each about 8cm (3in) long and 8mm (⅜in) in diameter. Lay horizontally on compost surface and lightly cover. Root in a frost-free greenhouse.
◐–◐◊ ✲✲✲

Choisya
Easily propagated shrubs. Take greenwood cuttings in midsummer and root in a propagator with bottom heat of 15°C (59°F). Take evergreen hardwood cuttings from late autumn to late winter, 20–25cm (8–10in) or 10cm (4in) long. Root in a frost-free greenhouse with humidity. Pot when rooted. In mild areas root under cloches; transplant the following autumn.
◐◊ ✤✤–✲✲✲

Cistus
These shrubs are easily raised from semi-ripe cuttings taken from midsummer to autumn. They can be rooted in a cold frame. Expect good results from cuttings taken in mid-autumn and rooted under a tunnel cloche, especially in mild areas. Transplant the following autumn when plants are well rooted.
◐◊ ✤✤

Clematis
These climbers are easily raised from leaf-bud cuttings taken from spring to late summer. Prepare the cuttings from soft or semi-ripe shoots. Root in warm humid conditions with bottom heat of 15°C (59°F). Semi-ripe cuttings need less humidity. Pot when well rooted (spring

COTONEASTER FRIGIDUS 'CORNUBIA' ♀

for semi-ripe cuttings). Keep young plants in sheltered conditions.
◐◊ ✤–✲✲✲

Cornus
(Dogwood)
These shrubs, grown for their coloured stems, are easily propagated from hardwood cuttings (20cm/8in) taken late autumn to midwinter. Root in a sheltered site, ideally in cold frames. If necessary keep well watered during the growing season. The frames can be opened fully in the spring. Transplant rooted cuttings the following autumn.
◐◊–◊ ✲✲✲

Cotoneaster
These shrubs are easily propagated from cuttings. Take softwood or greenwood cuttings from spring to mid-summer and root them in a propagating case with bottom heat of 15°C (59°F). Pot as soon as rooted. Alternatively take semi-ripe cuttings from mid-summer to autumn and root in cold frames or, in very mild areas, under cloches.
◐◊ ✲✲✲

Cytisus
These shrubs are moderately easy to propagate. Take semi-ripe cuttings in late summer or early autumn, with or without a heel. Use free-draining compost or rockwool modules. Providing humidity and bottom heat of 12–15°C (54–59°F) speeds rooting.
◎◊ ✻–✻✻✻

Daphne
Moderately easy shrubs to propagate. Take stem-tip greenwood cuttings from spring to early summer, or stem-tip semi-ripe cuttings in summer. Provide bottom heat of 15°C (59°F) for both.
◎–◐◊ ✻✻–✻✻✻

Elaeagnus
Moderately easy shrubs to propagate. Take semi-ripe cuttings from late summer to autumn. Wound at the base 2cm (¾in). Bottom heat at 15–20°C (59–68°F) speeds rooting. Alternatively take hardwood cuttings of vigorous growth from late autumn to late winter. Root in frost-free, humid conditions.
◎–◐◊ ✻✻✻

ERICA CARNEA
'VIVELLII' ♀

FATSIA JAPONICA ♀

Erica
(Heath)
Most of these shrubs require acid soil. Easily propagated from semi-ripe cuttings of non-flowering side shoots in summer. Insert in module trays, root in a propagator at 15–21°C (59–70°F). Pot when well rooted.
◎◊ ✻–✻✻✻

Escallonia
Easy shrubs to propagate. Take greenwood or semi-ripe cuttings from midsummer to autumn. Root greenwood cuttings with bottom heat, 15°C (59°F), semi-ripe in a cold frame. Root evergreen hardwood cuttings, 20–25cm (8–10in) long, from late autumn to late winter, in a frost-free environment.
◎◊ ✻–✻✻✻

Euonymus
(Spindle tree)
Easy shrubs to propagate. Take softwood or semi-ripe cuttings, in spring to late summer; greenwood cuttings, in late spring, for *Euonymus alatus* ♀; with bottom heat of 15°C (59°F).
◎–◐◊ ✻✻–✻✻✻

Fatsia
This shrub is easily raised from cuttings. Take semi-ripe tip cuttings 8–10cm (3–4in) in length, at any time. Remove all but the top two leaves. Insert cuttings so that only the bottom leaf joints are buried. Root in a propagator with bottom heat of 15–20°C (59–68°F).
◎–◐◊ ✻–✻✻

Fuchsia
Easily propagated shrubs. Take softwood cuttings at any time: nodal stem-tip, single-node and internodal stem cuttings all root readily. Take semi-ripe cuttings from midsummer to early autumn. Root in a propagator with bottom heat of 15°C (59°F).
◎–◐◊❀ (min.5°C/41°F)–✻✻✻

Grevillea
Moderately easy shrubs to propagate. They need acid soil, low in phosphorus. Take semi-ripe heeled cuttings from late summer into autumn. Best results gained with provision of bottom heat of around 15°C (59°F).
◎◊❀ (min. 5°C/41°F)–✻✻

HEBE × FRANCISCANA
'VARIEGATA'

SHRUBS AND CLIMBERS • 67

HEDERA HELIX 'GOLDHEART' (SYN. 'ORO DI BOGLIASCO')

Hebe
Easily raised shrubs. Take softwood stem-tip cuttings, spring to autumn, for large-leaved species. Provide bottom heat of 15°C (59°F). Semi-ripe stem-tip cuttings for all, midsummer to late autumn.
◐ – ◕ ◊ ✳ – ✳✳

Hedera
(Ivy)
Easily raised shrubs and climbers. Take leaf-bud cuttings of soft or semi-ripe shoots from spring to autumn. For bushy plants use adult growth. Keep shaded.
◐ – ◕ ◊ ✳ – ✳✳✳

Hydrangea
Easily raised shrubs. Root softwood stem-tip cuttings from late spring to mid-summer, with bottom heat of 15°C (59°F). Rooting takes 2–4 weeks. Root semi-ripe cuttings in midsummer, in a cool, frost-free place.
◐ – ◕ ◊ ✳✳ – ✳✳✳

Hypericum
Easily raised shrubs. Softwood or semi-ripe cuttings from spring to autumn, softwood cuttings in a propagator with heat, semi-ripe in a cold frame. Hardwood cuttings in pots from late autumn to mid-winter, in a cool greenhouse.
◐ – ◕ ◊ ✳✳ – ✳✳✳

Jasminum
(Jasmine)
Easily raised shrubs and climbers. Softwood or semi-ripe cuttings in spring and summer, in a heated propagator and cold frame respectively. Hardwood cuttings (20cm/8in) in winter for *J. nudiflorum* ♀ and *J. officinale* ♀, in a cold frame.
◐ – ◕ ◊ ♦ (min. 10–13°C/50–55°F) – ✳✳✳

Lavandula
(Lavender)
Moderately easy shrubs to grow from semi-ripe cuttings with a heel in summer or early autumn. Root in a cold frame or under cloches.
◕ ◊ ✳ – ✳✳✳

Lonicera
(Honeysuckle)
Shrubs and climbers are easily grown from cuttings: softwood or semi-ripe cuttings rooted with bottom heat from late spring to late summer; hardwood cuttings in cold frame from late autumn.
◐ – ◕ ◊ ✳ – ✳✳✳

Mahonia
Easily propagated shrubs. Prepare leaf-bud cuttings with 2.5–5cm (1–2in) of stem, from current year's shoots in autumn. Wound the base, reduce the leaves, and root with humidity and bottom heat of 15–20°C (59–68°F).
◐ – ◕ – ◕ ◊ ✳✳ – ✳✳✳

LAVANDULA ANGUSTIFOLIA 'TWICKEL PURPLE'

Passiflora
(Passion flower)
These climbers are easily propagated from leaf-bud cuttings taken from soft or semi-ripe shoots, spring to late summer. Root in module trays in a humid environment with bottom heat of 20°C (68°F). Overwinter under glass.
◐ – ◕ ◊ ♦ (min. 5–16°C/41–61°F) – ✳✳✳

Philadelphus
(Mock orange)
Easily raised shrubs. Take softwood cuttings from late spring and root in a propagating case with bottom heat of 15°C (59°F). Semi-ripe cuttings (8cm/3in) in summer will root in a cold frame.
◐ – ◕ ◊ ✳ – ✳✳✳

Philodendron
Shrubs or climbers, often epiphytic, easily propagated at any time from leaf-bud, stem-tip and stem cuttings, up to 10cm (4in) long, of soft or semi-ripe wood. Rooting takes 4–6 weeks in humid conditions at 21–25°C (70–77°F).
◕ ◊ ♦ (min. 15°C/59°F)

PYRACANTHA
'ORANGE GLOW' ♀

Pittosporum
These shrubs are moderately easy from semi-ripe cuttings in autumn. Place a 2cm (¾in) layer of sharp sand over compost. Rooting takes 8–12 weeks at 12–20°C (54–68°F).
☐–☒◊❀ (min. 5°C/41°F)–❀❀

Pyracantha
(Firethorn)
These shrubs are easily propagated from semi-ripe cuttings from late summer to early autumn. Root in a cold frame or, for quicker results, in a propagator with bottom heat of 15°C (59°F).
☐–☒◊ ❀❀–❀❀❀

Rhododendron
Acid-loving shrubs. Fairly easy. Softwood cuttings in spring for deciduous azaleas, in a humid propagator at 20°C (68°F). Semi-ripe cuttings midsummer to autumn for evergreens (same conditions). Wound bases.
☒◊–◊ ❀❀–❀❀❀

Ribes
(Currants and gooseberries)
Ornamental and fruiting shrubs. Easy from hardwood cuttings (20cm/8in) in late autumn or winter. Insert outdoors, but in a cold frame for ornamentals.
☐◊ ❀❀–❀❀❀

Rosa
Shrubs and climbers. Many roses are easily propagated from hardwood cuttings (20cm/8in) taken in winter, inserted in a sheltered spot outdoors.
☐◊ ❀❀–❀❀❀

Salvia
Shrubby salvias are easily propagated from semi-ripe cuttings taken from side shoots in late summer or early autumn. Root in a cold frame or cool greenhouse.
☐◊❀ (min. 7°C/45°F)–❀❀❀

Syringa
(Lilac)
Shrubs, moderately easy from softwood cuttings in spring. Root in a propagator with bottom heat of 15°C (59°F).
☐◊ ❀❀❀

Thymus
(Thyme)
These shrubs are easy from 5–8cm (2–3in) softwood cuttings in late spring or summer. Or take 5cm (2in) heel cuttings in late spring. Root in module trays with bottom heat of 15°C (59°F).
☐◊ ❀❀–❀❀❀

Viburnum
Shrubs. Evergreens are fairly easy from semi-ripe cuttings, late summer or autumn, with some bottom heat. Deciduous species from hardwood cuttings (20cm/8in) in winter, placed in a cold frame.
☐–☒◊ ❀❀–❀❀❀

Vitis
(Grape and ornamental vines)
Climbers, easily grown from hardwood cuttings taken in winter, with bottom heat of 21°C (70°F). Root standard hardwood cuttings in deep pots. Alternatively, prepare vine eyes consisting of a single bud and insert them in module trays.
☐–☒◊ ❀❀❀

Wisteria
These climbers are easily propagated. Softwood cuttings in late spring to midsummer with bottom heat of 15°C (59°F). Hardwood cuttings (20cm/8in) in winter in a frost-free greenhouse. Root them with bottom heat.
☐–☒◊ ❀❀❀

Yucca
Shrubs. Easy from leafless stem sections. Prepare 10cm (4in) cuttings from a mature stem. Insert single cuttings vertically into 9cm (3½in) pots. Keep humid at 21–24°C (70–75°F).
☐◊❀ (min. 7–10°C/45–50°F)–❀❀❀

VITIS COIGNETIAE ♀

PERENNIALS

NON-WOODY PLANTS, OR PERENNIALS, form a group of enormous value. They include hardy border plants, rock plants or alpines, aquatics and, in frost-prone climates, tender plants for summer bedding and greenhouse display. While division is a major method of propagation, many are easily raised from cuttings.

Acanthus
(Bear's breeches)
Architectural plants, easily propagated from root cuttings taken during mid- to late autumn. Lift the entire plant and wash roots free of soil. Choose roots of medium thickness. Cut into 5–10cm (2–4in) sections, making thinner cuttings the longest. Insert vertically in pots and cover with 1cm (½in) layer of grit. Root in a cold frame or propagator. Pot in spring.

Achillea
(Yarrow)
Easily propagated herbaceous plants. Basal stem cuttings taken in spring will flower within a year. Root in a propagating case with gentle bottom heat of 15–18°C (59–64°F). Semi-ripe stem-tip cuttings taken in early autumn can be rooted in a cool greenhouse or cold frame, potted into small pots when well rooted and overwintered in a cold frame.

Anemone
(Windflower)
Root cuttings offer the easiest means of increasing clumps of Japanese anemones. Take cuttings in autumn or winter by lifting the edge of a clump – do not take the entire plant. Root them in a cold frame and pot up when rooted. They usually flower in 2–3 years.

Artemisia
(Wormwood)
These herbaceous or woody based perennials are fairly easy to grow from cuttings. Take stem tips or heeled sideshoots as greenwood cuttings in late summer. (*A. absinthium* 'Lambrook Silver' ♀ roots best from softwood cuttings in spring.) Root in a propagating case with bottom heat of 18–21°C (64–70°F).

Aster
(Michaelmas daisy)
Easily propagated herbaceous perennials. Take basal stem cuttings in spring and root in a propagating case or under mist with bottom heat of 18–21°C (64–70°F). Pot and grow on in a cold frame.

Aubrieta
(Aubretia)
Easy-to-grow mat- or mound-forming plants. Take fully ripe cuttings in late summer or early autumn when shoots are mature. Root in a cold frame.

Begonia
A diverse genus of perennials. Stem-tip cuttings of stem-forming species, in spring or autumn, are easy, as are leaf cuttings in spring or summer of *B. rex*, *B. masoniana* ♀ and similar types. Use whole leaves or cut leaf into squares. Root in humid conditions at 21°C (70°F). (min. 5–13°C/41–55°F)

ARTEMISIA LUDOVICIANA

ASTER AMELLUS 'KING GEORGE' ♀

BERGENIA 'SUNNINGDALE'

DELPHINIUM 'BRUCE' ♀

Bergenia
(Elephant's ears)
These rhizomatous perennials are easy from leafless rhizome cuttings taken from mature plants in autumn. Wash the rhizomes free of soil and trim back any long roots. Cut rhizome into 4–5cm (1½–2in) sections, each with several dormant buds, half bury in moist perlite, buds uppermost, and root in humid conditions with bottom heat of 21°C (70°F). Pot up when rooted and grow on in a cold frame. Flowers in 1–2 years.

Campanula
(Bellflower)
Herbaceous plants, some good for rock gardens, moderately easy to propagate from cuttings. Basal stem cuttings of rock-garden types taken in late spring root in 2–3 weeks in a cold frame. Take stem-tip cuttings of herbaceous border species from new growth after flowering and root in a propagator with bottom heat of 18–21°C (64–70°F).

Chrysanthemum
Florists' chrysanthemums are easily propagated from basal stem cuttings in spring, generally taken from stock plants wintered under glass, or direct from garden plants. Root in propagator with bottom heat of 10°C (50°F). Pot when rooted, grow on in a heated greenhouse and plant out or pot on in late spring.

Dahlia
These tuberous perennials are easily propagated in spring from stock plants wintered in the dormant state under glass. Start plants into growth in early spring by increasing warmth and moisture and take basal stem cuttings when ready. Root at about 19°C (66°F) in humid conditions. Pot up, harden off in a cold frame. Plant out after frosts.

Delphinium
These border perennials are easily propagated from basal stem cuttings taken in late spring. Root them in pots of perlite (stand pot in a saucer of water) at 15°C (59°F) and pot as soon as rooted. Alternatively, root in cuttings compost. Harden off and plant out in early summer.

Dendrobium
(Orchid)
Take leafless stem cuttings in spring. Remove a 25cm (10in) long section of stem (cut just above a leaf joint). Divide it into 8cm (3in) long sections, each with at least one joint, cutting between nodes. Lay cuttings on moist sphagnum moss, cover, and root in warm, humid conditions. Cuttings should produce plantlets in a few weeks.

Dianthus
(Carnation, pink)
The perennial members of this genus range from hardy border and rock plants to half hardy perpetual-flowering carnations. All are easily propagated from semi-ripe cuttings, known as pipings, from mid- to late summer. Just snap them off. Root them in a cold frame, or in a propagating case with bottom heat of 15°C (59°F).

Diascia
These perennials are easily raised from softwood stem-tip cuttings in spring, or from the regrowth on plants trimmed after flowering in late summer. Root in a propagator with bottom heat of 18–21°C (64–70°F). Overwinter late-rooted cuttings in a cool greenhouse and plant out the following spring.

PERENNIALS • 71

Dieffenbachia 'Exotica'

Dieffenbachia
(Dumb cane)
Evergreen perennial, fairly easy from leafless stem sections in spring. Cut stem into 5cm (2in) pieces, cutting each below a leaf joint. Press flat into cuttings compost, buds uppermost, and root in humid conditions at 21°C (70°F). NB Sap is toxic.
◨◊❂ (min. 15°C/59°F)

Epimedium
(Barrenwort)
These rhizomatous woodland-garden perennials are easy to grow from rhizome cuttings

Eryngium alpinum

Euphorbia polychroma

taken in winter. Lift a clump and wash off the soil. Cut off the old leaves. Separate individual rhizomes, cut them into 5–8cm (2–3in) sections and treat them as root cuttings. Lay on compost surface, cover with 12mm (½in) of compost and place in a cold frame. Pot up when growth starts in spring.
◨◊ ✻✻✻

Eryngium
(Sea holly)
Border perennials, moderately easy from root cuttings in late autumn. Take cuttings from thick roots; cut into 5–8cm (2–3in) pieces. Lay flat on trays of compost, cover with compost and root in a frost-free greenhouse. Pot singly in spring when shoots appear. Harden off in a cold frame, plant out when established.
◨◊ ✻✻–✻✻✻

Euphorbia
(Spurge)
The border perennials in this varied genus are moderately easy from summer or autumn cuttings. Take stem-tip cuttings from mature growth after flowering, making them 5–10cm (2–4in) long. Allow to stand for an hour for the sap to dry. Root in a cold frame. Excess humidity can cause rot. Pot when rooted; plant out in spring.
◨–◨◊–◊ ✻✻✻

Geranium
(Cranesbill)
These herbaceous perennials are easy from root cuttings, especially *G. pratense*, *G. phaeum* and *G. sanguineum*. Take 2.5cm (1in) long cuttings in autumn and root in a cold frame. Root basal stem cuttings in spring at 15°C (59°F).
◨–◨◊ ✻✻✻

Impatiens
(Balsam, Busy Lizzie)
These easily propagated perennials are often used as greenhouse pot plants and summer bedding plants in frost-prone climates. Take soft stem-tip cuttings in spring or summer and root in humid conditions at 18–21°C (64–70°F). They will even root in water.
◨◊❂ (min. 5–10°C/41–50°F)

Impatiens New Guinea Group

Lewisia
(Bitterroot)
Perennials needing acid to neutral soil. Rosette cuttings in summer are moderately easy. Remove each offset with some stem and root in a shaded propagator or cold frame. Leaf cuttings in summer are more challenging. Use whole leaves and root in the same conditions.
◐–◓◊ ❋❋❋

Lobelia
Take stem-tip or stem cuttings from border perennials in summer. Moderately easy to root with bottom heat of 18°C (64°F). Flowering stems of *Lobelia siphilitica*, *L. cardinalis* ♀ and similar can be cut into 5cm (2in) lengths; remove lower leaves. For more plants, split cuttings vertically, retaining leaves on each half. Protect plants from cold over the winter months.
◐–◓◊ ❋❋ – ❋❋❋

Lupinus
(Lupin)
These hardy perennials are moderately easy from basal stem cuttings in spring. At a

LOBELIA × GERARDII 'VEDRARIENSIS'

LUPINUS 'CHATELAINE'

temperature of 15°C (59°F), rooting takes 10–14 days. To avoid rotting, insert cuttings in perlite. Pot when rooted and grow on in a cold frame or under cloches. Plant young lupins out in early summer in slightly acid soil.
◐–◓◊ ❋❋❋

Mentha
(Mint)
This herb is moderately easy from softwood cuttings in summer, rooted with bottom heat of 20°C (68°F). Easier are 4–8cm (1½–3in) rhizome cuttings in spring. Insert vertically and cover with 5mm (¼in) of compost. Keep at 10°C (50°F). Pot on singly when rooted and keep warm until established, then harden off and plant out.
◐–◓◊ ❋❋❋

Monarda
(Bergamot)
These border perennials are easy to grow from basal stem cuttings taken in spring and given bottom heat of 18–21°C (64–70°F). Alternatively, take soft stem-tip cuttings in spring and root in same conditions.

Pot on as soon as they have rooted and allow to establish before hardening off in a cold frame prior to planting out.
◐–◓◊ ❋❋❋

Nymphaea
(Water lily)
Easily grown from root-bud cuttings (rounded swellings with shoots on the roots) in spring or early summer. Cut out a bud, fill a 10cm (4in) aquatic basket with aquatic compost and press in the bud so tip is still visible. Immerse so bud is just below water level. Winter in a frost-free greenhouse.
◐❈ (min. 5–10°C/41–50°F)– ❋❋❋

Oenothera
Easily propagated herbaceous perennials. Take softwood tip cuttings or basal stem cuttings, especially of tap-rooted species which resent root disturbance, in spring and root them in a heated propagating case with bottom heat of 18–21°C (64–70°F). Once they have rooted, pot individually and allow to become established under

MONARDA 'CAMBRIDGE SCARLET' ♀

Osteospermum

These plants are generally used for summer bedding in frost-prone climates and propagated annually. They are easily grown from softwood cuttings in spring with bottom heat of 18–21°C (64–70°F). However, for bedding they are usually propagated from semi-ripe cuttings in late summer, with bottom heat or in a cold frame, and wintered under glass.
◐ ◊ ❀ (min. 2°C/36°F) – ❀ ❀ ❀

Papaver
(Poppy)
Oriental poppies (*Papaver orientalis* cultivars) are easily grown from root cuttings in late autumn. Take cuttings about 8cm (3in) long and insert vertically into cuttings compost. Keep in a cold frame or under cloches over winter. When well rooted in spring, line out in a nursery bed, or pot singly.
◐ ◊ ❀ ❀ ❀

PHLOX PANICULATA 'EVENTIDE' ♥

PENSTEMON 'SOUR GRAPES' ♥

Pelargonium
Zonal, regal, and ivy- and scented-leaved pelargoniums are easily raised from semi-ripe cuttings taken in late summer and given bottom heat of 15°C (59°F), or rooted in a cold frame, and wintered under glass.
◐ ◊ ❀ (min. 2°C/36°F)

Penstemon
(Beard tongue)
The border perennials and rock-garden types are easily raised from semi-ripe stem-tip cuttings taken in late summer or early autumn. The cuttings should root in two weeks at a temperature of 15°C (59°F). Protect rooted cuttings from frost in their first winter by keeping them in a cool greenhouse or cold frame.
◐ ◊ ❀ – ❀ ❀ ❀

Phlox
These plants are easily propagated from cuttings. Increase rock garden and woodland species by taking basal stem cuttings in spring. Alternatively, take softwood stem-tip cuttings in spring. Root both kinds of cutting at

PRIMULA DENTICULATA VAR. *ALBA*

15°C (59°F). Raise border phlox (*Phlox paniculata*) by taking root cuttings in autumn. Aerial parts of border phlox are prone to eelworm infestation, but roots are free of this pest. Lay the 2.5cm (1in) cuttings horizontally and cover with 12mm (½in) of compost. Root cuttings in a cold frame.
◐ – ◑ ◊ ❀ ❀ ❀

Potamogeton
(Curled pond weed)
These aquatic perennials are cultivated as oxygenators in garden pools, particularly *Potamogeton crispus*, and are easy to propagate. Regularly replace this fast-growing plant with young stock propagated from cuttings. Take stem-tip cuttings in late spring or early summer. Tie them loosely in bunches of six and plant in baskets of aquatic-plant compost.
◐ – ◑ ❀ ❀ ❀

Primula
(Drumstick primrose)
Primula denticulata ♥ and its colour forms are moderately easily grown from root

Rudbeckia 'Goldsturm'

cuttings in winter. Cut thicker roots into 4–5cm (1½–2in) pieces. Lay them flat on the surface of compost and cover with 12mm (½in) of compost. Root in a cold frame. When well rooted in spring, pot singly or line out in a shady nursery bed to grow on. They should flower in two years.

Rudbeckia
(Coneflower)
These border perennials are easily propagated from basal stem cuttings, taken as soon as they become available in spring. Insert in pots or module trays and root them in a propagating case with bottom heat of 18–21°C (64–70°F). When rooted, pot individually, allow to become established, then harden off thoroughly in a cold frame prior to planting out. The young plants may flower in the following year.

Saintpaulia
(African violet)
Grown as pot plants in frost-prone climates. Easy to propagate from leaf cuttings in spring or at any time when plants are in growth. Take fully developed, new leaves with their stalks (petioles) as cuttings. Insert in pots, either singly or several around the edge. Root in humid conditions with bottom heat of 21°C (70°F). Pot plantlets individually when they are large enough to handle. (min. 18°C/64°F)

Salvia
(Sage)
These border perennials are easy to grow from basal stem cuttings in late spring. Root at 15°C (59°F) to flower in the same year. Also easy from soft and semi-ripe stem-tip cuttings from new, non-flowering growth in summer or early autumn. Root in the same way. (min. 5°C/41°F)–

Sansevieria
(Bowstring hemp)
Grown as pot plants in frost-prone climates. Moderately easy to propagate from leaf

Solenostemon Wizard Series

Verbascum 'Cotswold Queen'

cuttings taken at any time of year. (Offspring from variegated cultivars will lack variegation if they are raised from leaf cuttings.) Use newly mature leaves and cut horizontally into 5cm (2in) sections. Insert vertically to half their depth, lower edge downwards, in a tray of compost. Root with bottom heat of 21°C (70°F) and avoid humid conditions. (min. 13°C/55°F)

Solenostemon
(Coleus, flame nettle)
These perennials are grown as pot plants or summer bedding plants in frost-prone climates and are easily raised from cuttings, ideally annually as young plants have the most attractive foliage. Take softwood stem-tip cuttings from early spring to late summer. They root readily in a free-draining medium such as rockwool, or even simply in a jar of water on a warm windowsill. They should root in 10–14 days at 18°C (64°F). Pot on as soon as the cuttings have rooted. (min. 2°C/36°F)

PERENNIALS • 75

Streptocarpus
(Cape primrose)
These perennials are grown as pot plants in frost-prone climates. Easily raised from leaf cuttings from spring to autumn. Cut a mature leaf in half along the mid-rib (discard the mid-rib itself) and insert shallowly, cut side down (see p.43). Alternatively cut leaf into transverse sections or chevrons 2.5cm (5in) deep. Insert, basal end down, in a tray. Root with bottom heat of 18°C (64°F). Keep humid.
◨◊❦ (min. 10°C/50°F)

Verbascum
Short-lived border perennials that can easily be increased from root cuttings in late autumn. Lift a vigorous plant and cut the thicker roots into 5cm (2in) sections. Discard the parent plant. Lay the cuttings flat on a tray of compost and cover with 12mm (½in) of compost. Place in a cold frame or, in very cold climates, in a frost-free greenhouse. Pot individually when rooted in spring.
◨◊❋❋ – ❋❋❋

Viola
(Pansy, violet)
These short-lived perennials root well from 2.5–5cm (1–2in) stem-tip cuttings in spring, prepared from new shoots before they elongate and become hollow. They root within 14 days at 15°C (59°F). Pot singly when rooted. Alternatively, three weeks before taking autumn cuttings, cut back plants; take stem-tip cuttings from the regrowth. Keep the young plants frost-free over winter.
◨–◨◊❋❋❋

ROSETTE-FORMING ROCK PLANTS

Androsace
(Rock jasmine)
Short-stemmed rosette cuttings in late spring and summer for small, cushion-forming species. Insert in pots of finely ground pumice, or horticultural or silver sand. Only the short stem is buried. Root at 10–15°C (50–59°F) in an unheated propagator or cold frame; keep shaded. Gentle bottom heat speeds rooting.
◨◊ ❋❋❋

Dionysia
These alpines are very difficult to root, being prone to rotting off, especially species like *D. curviflora*, *D.* x *tapetodes* and *D. freitagii*. Take rosette cuttings as for androsace. Remove with 5–10mm (¼–½in) of stem. Use hormone rooting powder.
◨◊ ❋❋❋

Draba
(Whitlow grass)
Cushion-forming alpines propagated from rosette cuttings as for androsace. *D. rigida* var. *bryoides* and *D. mollissima* benefit from rooting in freely draining ground pumice to help prevent rotting off.
◨◊ ❋❋❋

Morisia
Morisia monanthos is a rosette-forming, tap-rooted alpine, grown from root cuttings in late autumn and winter (moderately easy). Insert cuttings vertically in pure sharp sand, cover with 1cm (½in) layer of grit and root them in a cold frame. Keep only slightly moist.
◨◊ ❋❋❋

Saxifraga
(Saxifrage)
This is a large and varied genus but the small cushion types, such as *S. cebennensis* ♥, *S. oppositifolia* and *S. poluniniana*, are propagated from rosette cuttings as for androsace, and are best inserted in ground pumice.
◨–◨◊ ❋❋❋

ANDROSACE VANDELLII

DRABA MOLLISSIMA

Cacti and Succulents

Taking cuttings offers a reliable way of increasing many succulent plants. Use semi-ripe or ripe material. Root the cuttings in a mix of two parts cactus compost and one part fine grit, with a top-dressing of fine grit to prevent cuttings from rotting. Insert shallowly to avoid rot (*see also p.42*).

Aeonium
Tall species easily propagated from large, callused, stem cuttings, in spring or early autumn. Cut stem 8–30cm (3–12in) below rosette. Pot individually, 5–8cm (2–3in) deep. Rosette cuttings for smaller species. Root with bottom heat of 21°C (70°F). ◐◊❀ (min. 5°C/41°F)

Aloe
Easily propagated from leaf cuttings in spring and summer. Use whole leaves from halfway up the plant and set them in individual small pots. Root them in about 18°C (64°F). ◐◊❀ (min. 10°C/50°F)

Cephalocereus
(Old man cactus)
Moderately easy from cuttings, spring or summer, but only to save a plant that has rotted at the base. Take a large stem cutting from top of plant, allow it to callus and insert in an individual pot. Root with bottom heat of 18–24°C (64–75°F). ◐◊❀ (min. 10°C/50°F)

Cereus
These columnar cacti are moderately easy from stem cuttings in spring or summer. Cuttings up to 2m (6ft) long can be taken, but are generally shorter. Stand cutting on layer of fine gravel when potting. It may need support. Otherwise as for cephalocereus. ◐◊❀ (min. 5°C/41°F)

Crassula
Easy from 5–10cm (2–4in) stem cuttings in spring or summer. Moderately easy from leaf cuttings in spring or summer, but slow. Use entire leaves and insert upright. Allow both to callus and root with bottom heat of 18–21°C (64–70°F). ◐◊❀ (min. 5–10°C/41–50°F)

Echeveria
Often propagated from leaf cuttings in spring or summer. Moderately easy. Take whole leaves from main stem near base of rosette and insert upright after allowing to callus. Root with bottom heat of 18°C (64°F). ◐◊❀ (min. 7°C/45°F)

Echinopsis
Globular species and *E. chamaecereus* ♀ are easily propagated from the offsets which fall away. Use these as stem cuttings in spring or summer. Root with bottom heat of 21°C (70°F). ◐◊❀ (min. 5–10°C/41–50°F)

Epiphyllum
(Orchid cactus)
These epiphytic forest cacti are easy from flat stem cuttings taken from spring to late summer. Cut stems into 15–23cm (6–9in) lengths. Allow to callus, then insert upright. Root at 18–24°C (64–75°F) in humid conditions. ◐◊❀ (min. 10°C/50°F)

Euphorbia
(Spurge)
Stem-producing species such as *E. milii* ♀ and *E. canariensis* are moderately easy from well-callused stem cuttings, mid-spring to summer. Root with bottom heat of 21°C (70°F). ◐◊❀ (min. 7–15°C/45–59°F)

Haworthia
Moderately easy from cuttings, spring to autumn. Use severed offsets of taller species as stem cuttings. Leaf cuttings (whole leaves) for non-offsetting plants, but these are very slow to root. Root all with bottom heat of 18–21°C (64–70°F). ◐–◐◊❀ (min. 10°C/50°F)

Aloe aristata ♀

Hoya
(Wax flower)
Moderately easy to propagate from stem cuttings taken in spring or summer. Cut a length of stem just below a node and 3–4 nodes long. Dip base in hormone rooting powder. Root with bottom heat of 21°C (70°F). Rooting takes 2–6 weeks.
◱◊✿ (min. 7–13°C/45–55°F)

Kalanchoe
Easily propagated from stem or leaf cuttings in spring or summer. Take stem cuttings after flowering. Use whole leaves complete with stalk (petiole) and insert upright. Allow all cuttings to callus for 24 hours before inserting them. Root with bottom heat of 18–21°C (64–70°F).
◱–◱◊✿ (min. 10–12°C/50–54°F)

Lithops
(Living stones)
Cuttings in early summer are moderately easy to grow. Remove offsets of one or more heads from larger clumps and treat as globular stem cuttings. Allow to callus well. They are prone to rotting so be prepared for losses. Root with bottom heat of 21°C (70°F).
◱◊✿ (min. 12°C/54°F)

Mammillaria
(Pincushion cactus)
Moderately easy to propagate from spring or summer cuttings. Remove offsets and use them as globular stem cuttings. Some offsets can be pulled away easily, others will require a knife. Root with bottom heat of 21°C (70°F).
◱◊✿ (min. 7–10°C/45–50°F)

OPUNTIA MICRODASYS VAR. *ALBISPINA*

Opuntia
(Prickly pear)
These cacti are easy to grow from cuttings taken in spring or summer. Remove single pads as stem cuttings, cutting them off at a joint. Allow to callus for three days. Root with bottom heat of about 19°C (66°F).
◱◊✿ (min. 7–10°C/45–50°F)

Rebutia
These cacti are easy from cuttings in spring or summer. Sever offsets at their bases and use as globular stem cuttings. Allow to callus for 2–3 days, then root with bottom heat of 21°C (70°F).
◱◊✿ (min. 5–7°C/41–45°F)

Rhipsalis
(Mistletoe cactus)
These epiphytic forest cacti are easily propagated from stem cuttings in spring or summer. Detach a slender stem at a joint and cut it into 10–15cm (4–6in) long sections. Then treat as for epiphyllum cuttings. The cuttings should root in 3–6 weeks.
◱◊✿ (min. 10°C/50°F)

SCHLUMBERGERA 'SPECTABILE COCCINEUM'

Schlumbergera
(Christmas cactus)
Epiphytic forest cacti, easily raised from flat stem cuttings, 2–3 segments long, as plant starts growing. Treat as for epiphyllum cuttings. Root two back to back in a pot for larger plants.
◱◊✿ (min. 10°C/50°F)

Sedum
(Stonecrop)
Easy from cuttings in spring or summer. Tender plants from whole-leaf cuttings. Tender and hardy species from stem-tip or rosette cuttings. Leave to callus for a day. Leaf cuttings can be rooted on damp newspaper. Bottom heat of 21°C (70°F) for all.
◱◊✿ (min. 5°C/41°F)–✽✽✽

Sempervivum
(Houseleek)
Easy from cuttings in summer. Unrooted offsets can be severed at the base of their stem and used as rosette cuttings. Rosettes sit on compost surface with stem buried. Root with bottom heat of 21°C (70°F).
◱◊ ✽✽✽

INDEX

Page numbers in **_bold italics_** refer to illustrated information

A

Abelia 29
Acanthus 9, **_9_**, **_44_**, 69
Acer 30, 61, **_61_**
Achillea 69
adventitious roots 14, **_14–15_**
Aeonium 76
Ailanthus 29, 61
Aloe 76, **_76_**
alpines 46, 75
Androsace 75, **_75_**
Anemone 19, 69
aphids **_59_**
Argyranthemum 64, **_64_**
Artemisia 69, **_69_**
asparagus 45
Aster 38, 69, **_69_**
Aubrieta 69
Aucuba 32, 64
azaleas 16, 23

B

bark, compost mixes **_26_**
basal stem cuttings 29, 38–39, **_38–39_**
Begonia 9, 43, **_43_**, 69
Berberis 35, **_35_**, 64
Bergenia 45, **_45_**, 70, **_70_**
Betula 30, 61
blackleg **_55_**
blankets, propagating **_12_**
Bougainvillea 64
Buddleja 64

C

cacti 20–1, 46, 47, **_47_**, 76–77
Callistemon 64
Calluna 33, **_33_**
callus tissue **_15_**, 54
Camellia 17, 33, **_33_**, 40, **_40_**, 64
Campanula 70
Campsis 64
Caryopteris 64–65
Ceanothus 29, 35, **_35_**, 65
Cephalocereus 76
Cereus 47, 76
Chaenomeles 65
Chamaecyparis 17, **_34_**, 63
chamomile 18
chevron leaf cuttings 29
Choisya 65, **_65_**
chrysanthemums 38, 70
Cistus 65
Clematis 7, 9, 29, 40, **_40_**, 65
climbers 16, 23, 64–68
  hardwood cuttings 36
  leaf-bud cuttings 40–41
  semi-ripe cuttings 32
  softwood cuttings 30
cloches 11, 49, 52, **_52_**
clonal propagation 9
coir **_26_**
cold frames 11, 49, 52, 53, **_53_**, 57, **_57_**
Columnea 9
composts 26–27, **_26_**
conifers 16, 34, **_34_**, 63
Cordyline 46, 62, **_62_**
Cornus 62, 65
Cotoneaster 29, 65, **_65_**
Crassula 20, 42, 76
Cupressus 63
cutworms **_59_**
Cytisus 66

D

Daboecia 33
Dahlia 70
damping off 27, 55
Daphne 66
Delphinium 29, 39, 70, **_70_**
Dendrobium 70
Dianthus 19, 34, **_34_**, 70
Diascia 39, 70
Dieffenbachia 46, 71, **_71_**
Dionysia **_46_**, 75
diseases 13, 55, **_59_**
Draba 75, **_75_**
Dracaena fragrans 40

E

Echeveria 42, **_47_**, 76
Echinopsis 76
Elaeagnus 66
Elymus 45
Epimedium 71
Epiphyllum 21, 47, **_47_**, 76
Erica 33, 66, **_66_**
Eryngium 71, **_71_**
Escallonia 66
Euonymus 66
Euphorbia 71, **_71_**, 76
evergreens 32, 36

F

Fatsia 29, 66, **_66_**
Ficus 40, 62
florist's foam **_27_**
fog propagation 12, **_13_**, 56
Forsythia 36
frost damage **_59_**
fruit bushes 36
Fuchsia 6, 66
fungal diseases 49, 55, **_59_**
fungicides 28, **_28_**

G

Geranium 71
greenhouses 11, **_11_**, 48, 50, **_50–52_**, 54, **_54_**, 55
greenwood cuttings 30
Grevillea 66
grey mould **_55_**

H

hardening off 57
hardwood cuttings 9, 17, 23, 29, 36–37, **_36–37_**
Haworthia 76
heathers 33, **_33_**, 56, **_56_**
Hebe 23, 66, 67

Hedera **14**, 29, 40, **41**, 67, **67**
heel cuttings **15**, **29**, 35, **35**
herbs 45, **45**
hormone rooting powder 28, **28**
horseradish 45, **45**
hot beds 11, **11**
Hoya 9, 77
humidity 49, **50**, 54, 56
Hydrangea 9, **31**, 67
hygiene 25, **28**, 55
Hypericum 67

F
Ilex **17**, 62
Impatiens 27, 71, **71**
inert growing media 27, **27**
internodal cuttings 28, **28**, **33**, 40
Iris 45
iron deficiency **59**

J K
Jasminum 67
Juniperus 63
Kalanchoe 42, 77
knives 24, **24**
    sterilizing 25, **25**

L
Laburnum 37, 62, **62**
Lamium 9
Lavandula 67, **67**
leaf-bud cuttings **29**, 40–41, **40–41**
leaf cuttings 9, **9**, **29**, 42–43, **42–43**
leaf mould 26
leafless stem cuttings 46, **46**
Lewisia 72
Liquidambar 30, **30**
Lithops 77
loam 26
Lobelia 72, **72**
Lonicera **16**, 40, 41, 67
Lupinus 72, **72**

M
Magnolia 16, 23, 32, 62, **63**
Mahonia 9, **29**, 40, 67
mallet cuttings 35, **35**
Mammillaria **21**, 77
manganese deficiency **59**
Mentha **19**, 72
meristem 13, 14
Metasequoia 37, **37**
micropropagation 13, **13**
mint **19**, 29, 45, 72
mist propagation 12, **13**, **51**, 56
module trays 25, **25**
Monarda 72, **72**
Morisia 75
Morus 62

N
Nicotiana **13**
nodal cuttings **15**, 28, **28**, **33**, 40
Nothofagus 17
nursery beds 49, 58, **58**
Nymphaea 72

O
Oenothera **18**, 72–73
Opuntia 77, **77**
Osteospermum 73
outdoor cuttings 58–59

P
Pachyphytum oviferum **42**
Papaver 73
part-leaf cuttings **43**
Passiflora 40, 67
peat 26
peat blocks 27, **27**
Pelargonium 6, 73
Penstemon 27, 73, **73**
perennials 18–19, 23, 38–39, 69–75
perlite **26**, 39, **39**
pests 55, **59**
Philadelphus 30, 67
Philodendron 40, 67
Phlox 38, 73, **73**

Picea 63
pinching out 57, **57**
pipings **19**, 34, **34**
Pittosporum 68
plastic film 12
Podocarpus 63, **63**
Populus 37
Potamogeton 73
pots 24, **24**, 25
potting on 57, **57**
powdery mildew **59**
preparing cuttings 28–29
Primula **19**, 73–74, **73**
propagators **50**, **51**, 56
Prunus 30, 62–63
Pulsatilla 44
pumice 46
Pyracantha **28**, 68, **68**

R
Ramonda 9
Rebutia 77
regeneration 14–15
Rhipsalis 77
rhizome cuttings 45, **45**
Rhododendron 68
Rhus typhina 44
Ribes 68
Robinia 63
rockwool 27, **27**
rooting media 26–27, **26–27**
roots:
    regeneration 14–15, **15**
    root cuttings 9, **9**, **29**, 44–45, **44–45**
    rooting times 40
Rosa **30**, 68
rosemary 9, 29
rosette cuttings **29**, 46, **46**, 47
Rubus cockburnianus 40
Rudbeckia 74, **74**

S
Saintpaulia 9, 42, **42**, 74
Salix **29**, 63
Salvia 9, 29, 38, 68, 74

sand 26, 37, *37*
Sansevieria 29, 42, 74
Saxifraga 29, 75
Schlumbergera 47, 77, *77*
secateurs 24, *24*
Sedum 29, 77
semi-ripe cuttings 23, 29, 32–35, *32–35*
Sempervivum 77
shading 54, *54*, *58*
sharpening stones *24*
shoots, ripening 28
shrubs 16, 23, 64–68
  hardwood cuttings 36
  leaf-bud cuttings 40–41
  semi-ripe cuttings 32
  softwood cuttings 30
slugs 59
softwood cuttings 9, 23, 29, 30–31, *30–31*
soil, in compost mixes 27
soil-warming cables *51*
Solenostemon 9, 27, *27*, *57*, 74, *74*

stem cuttings 9, 23, 29
basal stem cuttings 29, 38–39, *38–39*
cacti, 47, *47*
leafless 46, *46*
stem-tip cuttings 29
stem roots 14, *14–15*
stock plants 23, *23*, 38, *38*
Streptocarpus 9, 29, 42, *43*, 75
succulents 20–1, 42, 46, 47, 76–77
suckers 44
Syringa 68

T
Taxus 27, 63
temperature 54, 57
"tenting" 49
Thymus 68
tools 24, *24*
transpiration 54
trays, module 25, *25*
trees 16–17, 23, 36, 37, 61–63
trimming cuttings 28
tunnel cloches *52*

U V
Ulmus 30
vegetative propagation 8–9
ventilation 49, 54, *54*
Verbascum 74, 75
vermiculite 26
Viburnum 68
vine weevils 59
vines 37, *37*
Viola 39, 75
virus diseases 59
Vitis 68, *68*

W Y
water, rooting in 27, *27*
water-retentive gel 27, *27*
watering 55, *55*, 59
whitefly 55, *55*
widgers 24, *24*
wind protection *58*
Wisteria 68
woody plants 16–17
"wound" roots 14, *15*
Yucca 29, 46, *46*, 68

ACKNOWLEDGEMENTS

Picture Research Anna Grapes
DK Picture Library Romaine Werblow
Index Hilary Bird

Dorling Kindersley would like to thank:
All staff at the RHS, in particular Barbara Haynes, Simon Maughan and Susanne Mitchell.

The Royal Horticultural Society
To learn more about the work of the Society, visit the RHS on the internet at
www.rhs.org.uk.

Photography
The publisher would like to thank the following for their kind permission to reproduce their photographs:
(key: t=top, b=bottom, r=right, l=left, c=centre, GPL=Garden Picture Library, MEPL= Mary Evans Picture Library, SPL=Science Photo Library)

2: GPL/Eric Crichton; **5:** GPL/Eric Crichton (bl); **6:** GPL/Paul Hart; **10:** MEPL; **11:** Peter Cross (r), GPL/John Glover (l); **13:** Holt/Nigel Cattlin (br), SPL/Sinclair Stammers (bc); **14:** GPL/Brian Carter (bl), SPL/Francoise Sauze (br); **17:** GPL/Lamontagne (bl, br), GPL/Ron Evans (tr); **19:** John Fielding (tl), GPL/Eric Crichton (b), GPL/J S Sira (tr), GPL/John Glover (c); **21:** GPL/J S Sira (t); **22:** Harry Smith; **48:** Holt/Rosie Mayer; **50:** Photos Horticultural (cl); **53:** GPL/Mel Watson (tl), GPL/Michael Howes (tr); **55:** Holt/Nigel Cattlin (bl), SPL/James King-Holmes (bc), Harry Smith (br); **59:** GPL/Vaughan Fleming (bl); Holt/Alan & Linda Detrick (br), Holt/Nigel Cattlin (cl, c, bc); **60:** GPL/Mayer/Le Scanff; **62:** GPL/Robert Estall (bc).

All other images © Dorling Kindersley.
For further information see:
www.dkimages.com